UNIVERSAL CONSTANTS IN PHYSICS

THE MCGRAW-HILL HORIZONS OF SCIENCE SERIES

The Science of Crystals, Françoise Balibar

The Dawn of Meaning, Boris Cyrulnik

The Realm of Molecules, Raymond Daudel

The Power of Mathematics, Moshé Flato

The Gene Civilization, François Gros

Life in the Universe, Jean Heidmann

Our Changing Climate, Robert Kandel

The Language of the Cell, Claude Kordon

The Chemistry of Life, Martin Olomucki

The Future of the Sun, Jean-Claude Pecker

How the Brain Evolved, Alain Prochiantz

Our Expanding Universe, Evry Schatzman

Earthquake Prediction, Haroun Tazieff

GILLES

COHEN-TANNOUDJI

UNIVERSAL CONSTANTS IN PHYSICS

McGraw-Hill, Inc.

New York St. Louis San Francisco Auckland Bogotá
Caracas Lisbon London Madrid Mexico
Milan Montreal New Delhi Paris
San Juan São Paulo Singapore
Sydney Tokyo Toronto

English Language Edition

Translated by Patricia Thickstun
in collaboration with
The Language Service, Inc.
Poughkeepsie, New York

Typography by AB Typesetting
Poughkeepsie, New York

Library of Congress Cataloging-in-Publication Data

Cohen-Tannoudji, Gilles.
[*Les Constantes universelles*. English]
Universal Constants in Physics / Gilles Cohen-Tannoudji.
p. cm. — (The McGraw-Hill HORIZONS OF SCIENCE series)
Translation of: *Les Constantes universelles*.
Includes bibliographical references.
ISBN 0-07-011651-2
1. Physical constants. 2. Physical measurements I. Title. II. Series.
QC39.C574 1993 92-21994
530.8'1—dc20

The original French language edition of this book
was published as *Les Constantes universelles*, copyright © 1991,
Hachette, Paris, France.
Questions de science series
Series editor, Dominique Lecourt

This book is printed on recycled, acid-free paper containing a minimum of 50% recycled de-inked fiber.

TABLE OF CONTENTS

INTRODUCTION

In Cambridge during May of 1687, Isaac Newton (1642–1727) wrote the preface to the first edition of *Principia mathematica philosophiae naturalis* [Mathematical Principles of Natural Philosophy], the book that would dominate the thinking of not only physicists but also all people of science and culture for more than two centuries. The author presented his work, written at the request of the astronomer Edmond Halley (1656–1742), with the intention of rigorously "establishing" and "demonstrating" the science he called "rational mechanics," (as opposed to "practical mechanics" designed to be used for machines), the mathematical study of "motion resulting from any forces and the forces that are required for any motion." This science, consisting of "general propositions" (definitions, axioms or laws, theorems, etc.), occupies the first two parts of the work; the third part presents an "example" to "explain the System of the World."

One day Newton, paraphrasing an old saying, declared "If I have seen further, it is by standing on the shoulders of Giants." In fact, he continued the work of Galileo and Descartes in mechanics and of Kepler in astronomy, and we know that he was influenced by the works of Christian Huygens. The first fifty years of the 17th century were not lacking in "giants!" The *Principia* may, in a certain sense, be read as the coronation of the most powerful movement of inventive thought demonstrated by humanity since antiquity. But Newton was not satisfied with merely amalgamating the results of others; starting with the announcement of his definitions and axioms, he began to innovate with an audacity that disturbed many of his contemporaries. Three of these innovations became the subjects of the most lively debates to accompany the progress of physics.

In contrast to the approach used by Descartes, he reintroduced the notion of "force" to science. Descartes, the author of *Principia philosophiae* (1644) believed that he had eliminated force as an obscure and confused idea linked to Aristotle's finalist metaphysics. He undertook to do so as a result of reducing matter to the extent of denying that there is any real difference between the interior and the exterior of a body. He thus presented a vast panorama of the world in which all motions occur through contacts, causing immense and incessant "vortices." Newton was opposed to these grandiose views, which appeared improper to a mathematical physicist, despite the stated ambition of Descartes. Newton redefined the quantity of matter as a "measure that can be taken both from its density and its volume," then the quantity of motion of a body as the "measure of both its velocity and of its quantity of matter." From that, he inferred that bodies move in space that is stationary and empty according to the trajectories that appear to be determined by forces which are themselves capable of being measured with all the mathematical precision desired.

Newton's second major innovation thus appeared immediately: The elaboration of a new method of calculation. In a memoir entitled *Method of fluxions and infinite series* (1671) he described a "rational method for the primary and the final reasons." It involved "infinitesimal" or, as we say today, "differential" calculus. Newton himself succinctly summarized the essentials of the conception of the mathematics forming the basis of this calculus: "I consider here the mathematical quantities, not only as being composed of the smallest possible parts, but also as described by continuous motion. Lines are described, and, having been described, are generated not by the apposition of parts, but by the continuous motion of points; surfaces by the motion of lines; solids by the motion of surfaces; angles by the rotation of sides; time by a continuous flux. These origins hold a real place in the nature of things and we observe them every day in the motion of bodies."

The application of "rational mechanics" conceived in this way was derived from the grandiose system that brought unparalleled glory to Newton. The statement of the law of universal gravitation—

the third innovation—uses the same equation of motion to describe "a stone being swung in a sling-shot" as well as "the planets revolving about other planets or heavenly bodies." Newton affirmed his hope that one day forces governing the motion of "particles making up a body" could be understood in the same way.

Once the persistent objections of the opposing Cartesians had been overcome, Newton's renown became universal; the respect accorded his work bordered on adoration. The philosophers of the Enlightenment considered themselves to be Newtonian in the realms of science as well as those of morality and politics. This book by Gilles Cohen-Tannoudji puts back into perspective the entire history of modern physics from the standpoint of modern research. Newton is the first of the four great figures on which his presentation is based. It is clear that, even after three centuries, physicists continue to base their explanations on his work. At the beginning of the 18th century it was acknowledged that the author of the *Principia,* by using a single mathematical formula to describe the forces governing the physical world, had succeeded in presenting a perfectly coherent if not complete design for them. He thus proved that human knowledge had found the path for mastering all of Nature. In this sense, Newton justified the hopes that had been placed in physics ever since the dawn of modern science.

Galileo himself was convinced that the "great book of Nature" would be written in the language of mathematics and Newton proved him right. In doing so, he not only succeeded in establishing the relationship of thought to the understanding of Nature, but also provided a representation of thought itself which had been the object of fierce debate.

A certain very harsh rationalist view of history holds that modern science is the result of a sudden conversion of minds to the "active life." The world at the end of the Middle Ages was torn from the torpors of contemplation to subject Nature to "the question," in the full meaning of the word intended by Lord Chancellor Bacon. But the "conversion" was long to develop, and it was preceded by a profound spiritual movement. This was the case, for example, of the fire-and-brimstone works of the Dominican Giordano Bruno (1548–1600). The German philosopher Ernst Cassirer found just the right words to express their essential tone: "desires and sensuous pleasure join together with the power of the spirit to tear humans away from simple reality and allow them to imagine what may be possible." The Inquisition burned Bruno at the stake on February 17, 1600, under the most horrible conditions. The ecclesiastic tribunal was perfectly aware of the danger: the crime of the rebellious Dominican lay not only in his adherence to the heliocentric theories of

Copernicus; he also undoubtedly made the mistake of believing in "natural magic," the hermetic tradition that divided the Church and the monarchies of the Counter-Reformation. In the eyes of his persecutors, was not his most serious mistake to have proudly based the message of a new world on this archaic doctrine? His proclaimed hypothesis of the infinity of the Universe and of the existence of other worlds appeared as the cursed symbol of a train of thought whose arrogance could be mistaken for demonic possession. Thus the bell tolled for the medieval conception of the relationship of thought to being. All reality until then had found its place in an immutable divine architecture; by the distance that separated it from the "primary cause," each being received an absolutely predetermined "value." Knowing, by the exercise of natural reason, was equivalent to perceiving this place and discovering this hierarchy. Nature appeared to be limited by a predetermined horizon, by the impenetrable frontier which, separating all creatures from the Creator, condemned them in advance to imperfection not only in their knowledge but also in their work. Bruno caused a veritable upheaval that threatened to destroy this edifice: Nature, according to him, fully participates, from within, in the original divine being, which is thus the "soul of the world." Consequently, the creative power of thought alone could assert itself as infinite!

Galileo undoubtedly shared some of this conviction; but he expressed it in completely different language by using an image that preserved the idea of a personal God maintaining a relationship of exteriority with creatures. God is the author of the "great book of Nature." Thus, we must accept the fact that two books were given to humanity. Galileo hinted that instead of the obscurities of the Word revealed in the Scriptures, one may prefer the clarity of "mathematical language" written by Nature because this language appears fully accessible to human understanding. However, he offered no explanation for the metaphysical abyss opened by his conception of these two books. Descartes, who admired him as a scholar, deplored the mediocrity of his philosophy. His particular genius was therefore applied to establishing the philosophical bases that would be appropriate to the new mechanics. The *Meditationes de prima philosophia* [Meditations on the First Philosophy] (1641) showed the way to this foundation: renewing the sense of the word "creation." Descartes presents God as having created the essence of things and "eternal truths" with which he "seeded" our minds; God thus appears as the "guarantor" of the way we apply these truths to Nature. All we have to do is to follow the correct method, the one analytic geometry teaches us, to make the world transparent to our minds through natural light. The "modern" thrust of thought, and the autonomy of its process are thereby assured. Nevertheless, Descartes avoided the pitfalls of Bruno's audacity: God, being an infinite being, remains incomprehensible to finite thinking which encounters an "ontological" limit.

The philosophers of the Enlightenment were critical of the Cartesian systematic approach, rejecting his assumptions and his metaphysical conclusions. Wanting to break the last connection remaining between the knowledge of Nature and some supernatural origin, they believed that they could enlist the work of Newton in this combat. Rather than attribute the progress of the human mind to the virtues of a supposed "natural light," they declared Nature itself to be illuminating. Jean Le Rond d'Alembert (1717–1783) asks the following question in his *Eléments de philosophie*: "What does it matter to us, in fact, to penetrate the essence of bodies (. . .) provided the general system of phenomena, always uniform and continuous, does not confront us with any contradiction?" We will not seek to base the unity of Nature on the unity of its divine origin; let us content ourselves with the knowledge that it presents a constant and complete order, totally intelligible as a perfectly closed system, a single and uniform entity in and of itself.

It was thus around the universality of the law of gravitation that the cult of Newton was organized.

This version of "Newtonianism" gave rise to a veritable epistemological tradition that began in France with the texts of the great astronomer and mathematician Pierre Simon Laplace (1749–1827), continued with the work of Auguste Comte, and has come down to us in various guises of philosophical positivism. Science, it is said, must renounce trying to determine the "why" of phenomena and study only the "how." Scientists must be content to develop mathematical formulations for the regularities that are common to facts established by observation. Such a conception hides a quandary of thought that did not escape Newton himself nor some of his contemporaries: How can the idea of Nature be deprived of the support provided until then by the idea of God without foregoing the means of explaining the universality, unity, and eternity of its laws? Especially if we want to avoid the solution proposed by Spinoza, which in some respects is similar to that of Bruno, namely a pure and simple identification of God with Nature? When Newton felt the need to write the *Scholium* in 1713 to add to the second edition of the *Principia*, he approached the question directly to clarify some misunderstandings. There is no "mechanical" cause for the order of the world, he explained. "This extraordinary arrangement of the Sun, the planets and the comets can only have originated in the design and domain of an intelligent and powerful being. This Being governs everything, not as the soul of the world, but rather as the lord of all that exists." Just as many others did, d'Alembert deliberately ignored this profession of faith, which harks back to a theological tradition foreign to him.

The Scottish philosopher David Hume (1711–1776) observed that when we apply our efforts of knowledge to the axiom of the uniformity of Nature, we are resorting to simple belief! Hume sought the basis for this belief in an analysis of "human nature." Later, Emmanuel Kant was to say that this argument awakened him from his "dogmatic sleep." The author of the "Theory of Heaven" wrote the *Critique of Pure Reason* (1781) to "rescue" the objectivity of Newtonian science that had been undermined by the radically skeptical reasoning of Hume. The destiny of Western philosophy was to be completely changed, because this salvation or this rescue could be achieved only by separating absolute space and time from the divinity of the *Principia* in order to record them in the structure of the human mind as "*a priori* forms of sensitivity."

Henceforth the question of the "limits" of knowledge would be formulated in the following terms: human beings would never know things in themselves; what would later be called human "finiteness" condemns us to be able to construct through the use of our faculties only "objects" about which we can say no more than that they "correspond" to the existence of these things because they are based on perceptible facts emanating from them. The objectivity of knowledge produced by physics thus appears as the horizon of a process of objectification regulated by an uninterrupted demand for unification which refers to the unity of the thinking person. From the standpoint of the history of science, the most important aspect of Kant's work resides in the sanction it thus gives to the Newtonian concept of the relationship between matter and space. Space, as a form of intuition, is not "of things;" it is a reality radically different from the matter that fills it without ever acting on it.

As prestigious as the author of this philosophical construct became, it would be difficult to understand the power it exercised over thinking until the end of the 19th century if the accumulated and coordinated successes of rational mechanics to which it referred were not acknowledged. Auguste Comte gave an encyclopedic picture of these successes in his *Cours de philosophie positive* [Positive philosophy course] written between 1830 and 1842: from "celestial mechanics" to the "analytical theory of heat" of Joseph Fourier (1768–1830), the same way of thinking in each case manifested an extraordinary productivity.

It is easy to understand how, under these circumstances, the seriousness of the confusion that gripped physicists when the second law of thermodynamics was rediscovered in 1850, as formulated by Sadi Carnot in 1824 in his memoir *Réflexions sur la puissance motrice du feu et sur les machines propres à developper cette puissance* [Reflections on the Motive Power of Fire and on Machines Fitted to Develop that Power], caused the appearance of an order of phenomena that are not encompassed anywhere in this construct. Although it is possible to completely transform work into heat, the contrary is not true. Carnot gave the theoretical formulation for this practical impossibility experienced by engineers in the development of steam engines. The Austrian physicist Ernst Mach (1838–1916) was to derive some philosophical lessons from it decades later in his famous book *Kultur und Mechanik* [Culture and mechanics]: the irreversibility of thermodynamic processes must lead to the re-examination of the hypotheses of "mechanism." Wanting to explain everything in terms of "mechanical models" is to submit to an outright philosophy of nature that has committed the initial error of absolutizing the schemes of thought of rational mechanics, which thus accorded excessive credit, following Kant's example, to the notions of absolute space and time.

Ludwig Boltzmann (1844–1906) attempted to discover the macroscopic laws of thermodynamics and to account for the irreversibility of these phenomena by interpreting them as the apparent effect of the interactions of a very large number of elementary systems composed of atoms and molecules. To accomplish this, he used probability calculations. The opposition of Mach and his followers was immediate, violent, and radical: they took this conjectured atomism for a new incarnation of the philosophy they were fighting. They rejected the idea of the atom, and were incapable of appreciating the true value of the statistical intuition of their adversary. An epistemological conviction explained by Mach himself in several works supported their position, namely that physicists should abstain from any hypothesis about the realities they study and just be content to establish a correspondence between mathematical formalism and observed facts. The "logical empiricism" of the Circle of Vienna (1929), originally created as the Ernst Mach Society, studied anew the essentials of these theses. In a highly coherent way, the anti-Kantism of Mach redirected his thinking to a position close to that of Hume's, the moment he rejected any recourse to "metaphysics."

Albert Einstein (1879–1955) more than once underscored his debt to the author of *Culture and Mechanics.* In this book he discovered the absurdity of the idea of immediate action at a distance accepted by Newton and the connection that unites this absurdity to the notions of absolute space and time. However, Einstein never had the least sympathy for the positivism of Mach. The reason for this was the admiration he professed for Newton himself. From Newton's work he distilled the idea that the vocation of physics is to construct a "unified theory of the universe" around the smallest possible number of mathematical formulae. Gilles Cohen-Tannoudji brilliantly explains the approach that allowed the inventor of the special theory of relativity (1905) and then the general theory of relativity (1915) to achieve this program on his own terms. Once the assumptions of rational mechanics were brought to light, Einstein recast them in a way of thinking that "comprises" them, in the sense that it encompasses them while mathematically accounting for their approximation. However, the reader will also see how, even before it was ever achieved, the unification thus predicted for physics was undermined by the "quantum revolution." How is it possible to understand that matter on a subatomic scale obeys different laws than on "our" macroscopic scale? Einstein never accepted it. But the philosophical question thus posed probably harks back one last time to Newton and to the ambition he communicated to generations of physicists to construct a "system of the world." God, this all-powerful being who is found nowhere but is omnipresent, assures the coherence of this system in the *Principia*; the knowledgeable "subject" holds his place in Kantian philosophy. We have seen what the cost of this has been to physics. In contrast to the intellectual resignation of positivism which restricts science to the recording of measurements made by researchers, it brought Einstein to refer back to Spinoza: it is the being of Nature which is revealed in the laws we formulate. More generally, in opposition to all theologies, Einstein relied on "cosmic religions": the momentum of knowledge is generated by the admiration that all of Nature inspires. Quantum mechanics demonstrates the interaction of the measuring devices with the objects being studied, and it obliges physicists who work on the atomic scale to integrate experimental conditions into conceptualization itself. Does this mean that the mind would be an intimate part of matter, and that modern physics would rejoin a certain religious view of the world to celebrate the sovereignty of the former over the latter? Some say so, those who are visibly more concerned about salvation by God than by the progress of science. Doesn't modern physics suggest we stop trying to take the place of God and leave science where it belongs? The processes of unification of physics never actually present us with anything but a sum of "points of view" on the world, dependent on the cut-out models that we must make of reality to know it. The universal constants, "cornerstones of theoretical physics," to use Max Planck's expression, mark with prodigious precision the limits of our effort in the direction of the infinitely small as well as the infinitely large.

For the scientific mind, writes Gaston Bachelard, "to clearly demarcate a frontier, is to already cross it." Gilles Cohen-Tannoudji, far from losing himself in finalistic speculation which believes it can discover the hand of God in the meticulous "regulation" of the constants, works the notion of "horizon" to the point of creating a veritable philosophical category. He draws attention to the singular way in which physicists play with the limits of their knowledge. He pleads for the humility of the scholar; but he also demonstrates how horizons, always receding, never cease to "inspire" the progress of knowledge. The reader will undoubtedly discover with passion the new line that he draws at the cutting edge of research in astrophysics and particle physics: the promise of a scientific cosmology. Tomorrow we may perhaps discover new constants of the same type, but will we not once again have to recast our conception of the cosmos?

Dominique LECOURT

I

G AND c: RELATIVITY

The terms "constants" and "universal" suggest eternal and all-encompassing truths. But the world appears to us only in terms of our relationship to it here and now. How can physics raise the result of its local and immediate observations to the level of eternal and universal truths? This book attempts to answer this very question.

The universal constants we will discuss are Newton's constant G, Boltzmann's constant k, the speed of light c, and Planck's constant h. These constants play a fundamental role in the structure of physics: its organization into autonomous disciplines when the constants are considered separately, but also its unification when two, three, or even all four of the constants are considered simultaneously. During its history, physics has been able to introduce other constants, such as the vacuum dielectric constant, or the Hubble constant, but I believe that only G, k, c, and h are universal constants. This opinion is not unanimously accepted by physicists, some of whom think that other constants are just as fundamental, or that not all of the four have the same importance. Such a diversity of opinions should not be surprising, because nothing compels physicists to unanimity on the subject of the epistemology of physics.

Essentially, the response that I propose to the question stated above resides in the hypothesis that the universal constants express the *inherent limitations of human knowledge.* The limits are not only inevitable and inalienable, but also capable of shifting, like *horizons*. I propose to demonstrate that the four universal constants are a manifestation of the existence of horizon lines that separate us from the infinitely small and the infinitely large. More precisely, I have divided these constants into two pairs, G and c, on the one hand, and h and k, on the other. G and c are the constants of relativity involved in our relationship to the Universe in its global aspect. Since they are linked to the relative character of the notion of simultaneity, it is impossible to define the same "now" everywhere. As for h and k, which can both be interpreted as *quanta*, they are encountered when the microscopic structure of matter is examined. These two constants delimit a temporal horizon; they reflect that what is "here" now certainly has not always been, and certainly will not always be.

Actually, the universal and eternal truth that these four constants express is that the Universe eternally exceeds our capacity to represent it.

Because old illusions of anthropocentrism and modern illusions of reductionism and scientism were dissipated by this humility, physics has become more clear-sighted as a result of the *quantum revolution* which has allowed great advances to be made in science and technology in the 20th century.

Obviously such an interpretation of universal constants did not appear all at once. But in the sciences, according to a keen observation by Gaston Bachelard, the old must be rethought in terms of the new; it is the new that gives meaning to the old by regenerating it.

THE LAW OF UNIVERSAL GRAVITATION AND NEWTON'S CONSTANT

Before all else, we should keep in mind the experimental character of the physical sciences, even if debates about the physical interpretation of their results and the development of their hypotheses are fascinating. In fact, the statement of the four constants turns out to be indebted to fundamental experiments. This was the case for Newton's constant *G*. Before this constant could appear in the framework of the universal theory of gravitation, many discoveries had to precede it. The experiments of Galileo had to establish that the acceleration imparted to a body by gravity is independent of the mass of the body. Patient and meticulous observations of the motion of planets around the Sun had to be made by Tycho Brahe (1546–1601). A phenomenological model had to be developed by the German astronomer Johannes Kepler who, from these observations, was able to formulate the famous laws that bear his name; we had to know that these bodies do not describe circles, but rather ellipses; that the planets sweep equal areas in equal times, and that the period of their revolutions is a function of the long axis of the ellipse. This history, which combines experiment, observation and interpretation, was truly the indispensable preliminary to the formulation of Newton's laws. Richard Feynman did explain that Newton was satisfied to assemble all these observations into a single theory without ultimately adding anything to what had been said previously but separately by Kepler and Galileo. What indeed is the essential idea of the "law of areas"? That the force that attracts a planet to the Sun also attracts the center of the Sun to the planet (it is said to be "central"). What does Galileo's law assert? That the acceleration due to gravity is independent of the mass. Newton established a previously unrecognized relationship between these two laws. The law of equal areas is easily explained if it is assumed that the planet experiences a constant acceleration directed toward the Sun; it is thus a consequence of the hypothesis of a central acceleration which is the inverse of the square of the distance. The converse can be deduced mathematically: this acceleration produces a motion whose trajectory is an ellipse. In modern terms, the force of gravitation is proportional to the product of the two masses of the interacting bodies and is inversely proportional to the square of the distance between them.

If we wish to express this law mathematically, we can use an equation in which a force corresponds to the product of two masses divided by the square of a length. Thus mass, length, and time interval designate fundamental physical quantities. Mass is defined as the amount of matter contained in a given volume, length as the amount of space between two points on a line, and time interval as the amount of time between two instants. All other physical quantities can be expressed in terms of the fundamental physical quantities from which they are derived. "Dimensional content" of a derived quantity represents the proportions of the fundamental quantities used to define it. For example, velocity has the dimensional content of length divided by time, and acceleration has the dimensional content of length divided by time squared, and so on. A physical law expresses a relation between physical quantities that have the same dimensional content. Now, specifically, if we consider the law of gravitation, we see that the dimensional content is different on each side of the equation! What, in fact, is the actual dimensional content of a force? It is, as Newton's first law teaches us, the product of mass and acceleration, hence, a mass multiplied by length divided by time squared. This is the left-hand side of the Newtonian equation for gravitation. When we consider the right-hand side, however, we see that it has mass squared, because the product of two masses is divided by length squared. Thus, there are no times or masses, etc. It was to eliminate this disequilibrium that the constant G was introduced. Although it has been attributed to Newton, it was never even discussed by him. It permits us to re-establish the homogeneity of the dimensional content of the left-hand and the right-hand sides of the equation that expresses Newton's law in modern mathematical language: $F = G\ Mm/r^2$ (m being the mass of the body, M the mass of the Earth and r the distance of the body from the center of the Earth).

The term "constant" indicates that this quantity must be independent of time, position, mass, and the nature of the body. If G were not a *universal constant*, then the *universal* law of gravitation would not be what it is: if G depended on time, the force of gravity would vary with time, and the planets would not always orbit in the same way; if G depended on the position of the body interacting with gravity, the law would not be inversely proportional to the square of the distance.

As we will see, such a constant, although it was not originally thought of as an essential characteristic of Nature, does at this stage of our knowledge have a "utilitarian" content. It appears exclusively as a quantity necessary for the correct mathematical formulation of the law—a quantity with a peculiar dimensional content (length cubed divided by a mass and time squared, or 6.67×10^{-11} m^3 kg^{-1} s^{-2}, to be precise), the significance of which we cannot precisely comprehend at this moment. In fact, at first glance, the introduction of this constant only seems to disturb the beauty, harmony, and elegance of the law that, since Newton's time, has amazed thinkers and given rise to the essential philosophical speculations. A physicist is reluctant to introduce a quantity that seems to require only calculation. Later on, toward the end of this book, when we make a connection between h and c, we will see that G acquires a highly surprising significance that is related, where subatomic realities are involved, to the microscopic structure of space and time.

SPECIAL RELATIVITY AND THE SPEED OF LIGHT CONSTANT c

Newton's work formed the basis of an impressive development, that of "rational mechanics," which took final shape in the work of the French mathematician Joseph Louis Lagrange (1736–1813) and the Irish mathematician William Rowan Hamilton (1805–1865). As the Austrian physicist Ernst Mach (1838–1916) pointed out, Newton's mechanics remain geometric, proceeding by a synthetic approach, in which hypotheses are deduced by using constructions and figures. Lagrange and Hamilton followed an "analytical" approach by applying algebra to geometry to determine the conditions for the existence of the properties of a figure. "Classical" physics, which depends on this development, thus appears as a systematic mathematization of all mechanics of material bodies. Then, in the very middle of the 19th century, the synthesis achieved by James Clerk Maxwell (1831–1879) appeared, combining electrical, magnetic, and optical phenomena in the same overall theoretical treatment. This admirable synthesis appeared as a new victory for physics! Thus, Maxwell came to defend a *wave* theory of light, whereas Newton had supported a *corpuscular* conception! It is interesting to compare the dynamics of the waves described by Maxwell's equations to the dynamics of material bodies described by the equations of Lagrange and Hamilton. It actually seems possible to unify the two dynamics, but serious problems arise in connection with the time necessary for the propagation of light. In fact, it does not appear to be entirely possible to apply the fundamental laws of rational mechanics to the propagation of light.

The experiment performed by Albert Abraham Michelson and Edward Williams Morley surfaced in this context in 1887. Repeated in 1902, and again in 1903, it was an extremely simple and clever interferometry experiment. The two American physicists attempted to determine whether interference phenomena between light beams coming from different sources are shifted in two different directions at different speeds. However, contrary to every expectation, such interference does not occur! The evidence was undeniable: the propagation of light is definitely a particular case that defies the laws of classical physics, and checkmates them. Only the special theory of relativity was to allow this difficulty to be overcome.

Henri Poincaré (1854–1912), the great French mathematician and physicist, who continued on a new basis the reasoning of Ernst Mach, observed that the idea of "immediate action at a distance," which Newton had accepted to explain the attraction of bodies in a supposedly empty absolute space, was not at all obvious or even very reasonable. If we assume, as Einstein resolutely did, that such an action does not exist, this means, roughly speaking, that if something is done at a point A, the effect of what is done takes a nonzero finite time to be transmitted to a point B located a finite nonzero distance away. But to speak of finite time is tantamount to saying that there can be no infinite speed. Nothing can propagate at an infinite speed. The conclusion is obvious: there must be a speed that represents the upper limit of all speeds, and this in the entire Universe. It also appears obvious that this speed must be *invariant*, namely, it must not change with changes in the "reference frame," a term that is used in physics to designate the coordinate system to which we "refer" to locate the motion we wish to study. If, in fact, such a variation proves to be possible, then we could exceed the speed limit, which is a contradiction!

However, nothing has ever been found with a speed greater than the speed of light. Moreover, the Michelson-Morley experiment demonstrated that the speed of light is independent of the frame of reference. It had to be deduced that the speed of light c, 300,000 km s^{-1}, can be taken as the upper limit of all speeds.

Einstein brought this impeccable reasoning to fruition in 1905, and extended it by re-examining which principles would be overturned in classical physics by his conclusions. It was thus that he retained certain principles and reformulated others. The special theory of relativity can be defined as the field of the redefinition of classical physics caused by the discovery of an upper speed limit affecting the entire Universe. Einstein retained the Galilean principle of relativity, which had allowed the author of *Dialogo sopra i due massimi sistemi del mondo* [Dialogue Concerning the Two Chief World Systems] (1632) to show that "the butterflies aboard a ship fly to and fro in the same manner, regardless of whether the ship is motionless at the dock or underway between Venice and Aleppo" and to conclude from this that if the Earth rotates, we do not perceive its motion. But Einstein liberated physics from the absolute character of the notion of "simultaneity" that had been accepted first by Galileo, and later by Newton. They both assumed that the simultaneity of two events that do not occur at the same place has an absolute sense, that it is true for every observer, wherever that observer is located. It then became necessary to reformulate the principle of relativity itself and to develop the first "theory" of relativity.

In classical physics, space and time are independent. Space has three dimensions, and time does not depend on the space reference frame: "now" can be defined everywhere the same way. It has been said that time is a "spatial scalar." For Einstein, on the contrary, time appeared as the fourth dimension of "space-time"; it is then relative to the space-time reference frame. It was thus appropriate to reformulate all of mechanics, including the principle of relativity. What appears as a point in space now becomes a discrete *event* in space-time, an event indicated by four numbers: the date and the three coordinates of the place where it occurred. It is obvious that we must redefine the transformations of uniform rectilinear motions, as conceived since the time of Galileo, by inserting them into the space-time frame, and not just into space. It is thus with respect to space-time transformations that we must now speak of relativity. Galilean relativity has thus been generalized. The price of this theoretical innovation is first of all that time, as a coordinate, no longer plays the role of an invariant. The same is true of the *space measure*. The length of a ruler is held to be invariant in classical physics, independent of the reference frame; but if, along with Einstein, we consider the three space components of a four-dimensional vector, this is no longer the case. Now the measure or "metric" of a vector connecting two points can change if the space-time reference frame is changed. The way in which time and the space measure change as a result of a space-time reference-frame change is adjusted so that the speed of light remains constant.

Einstein did not stop here with so fruitful an approach: even mass no longer appears as an invariant. Newton's law linking force, mass, and acceleration must in turn be reconsidered. And we can readily understand why. Let us begin, as Einstein suggested, with the statement of this law: when a force is exerted on a massive body, the acceleration acquired by this body is proportional to this force and inversely proportional to its mass. This Newtonian law is perfectly applicable, but only when the speed of the body is small compared to the speed of light. Since no body in the Universe is capable of exceeding this speed, when the speed of a body approaches the speed of light, the energy imparted to that body must of necessity stop causing an acceleration—by definition henceforth impossible. There is only one solution; namely, that this energy increases the mass of the body, so long as the upper speed limit is never exceeded. Einstein's well-known formula $E = mc^2$ (where E is the total energy of a body having a mass m) expresses this variation in mass. Mass is no longer invariant; it depends on the speed of the body.

The "invariant mass" of a body is defined as the mass of the body in a given reference frame or at rest. For particles which, like "photons" or "light particles" are never at rest, always moving at the speed of light, the invariant mass is zero. Photons can, however, have more or less energy; they then have an "effective mass" which, according to Einstein's equation, is equal to their energy divided by the speed of light squared.*

It is impossible not to admire the power of Einstein's reasoning. By logical deduction alone, he was

*Some authors, such as the Soviet physicist Lev Okun, to whom I am grateful for the suggestion he made to me, think that only invariant mass has a precise physical meaning and that use of the term "mass" should be avoided when designating any other "mass" whatsoever.

able to obtain results whose experimental implications have since been tested and verified countless times. For example, the fact that particles cannot exceed the speed of light when they are "accelerated" is verified on a daily basis in large accelerators. (The particles are not actually accelerated—they are simply given more energy.) The precision with which the beams of particles that circulate in these accelerators are manipulated confirms that they never exceed the speed of light.

We now return to the special theory of relativity itself. It allows us to determine that time, distance, and mass are not invariant, but that these physical variables are "relative" to the reference frame. There is, however, no trace of *relativism* in this statement; on the contrary, as Victor Weisskopf* pointed out, for the first time we have a theory that allows laws of physics to be formulated independently of the reference frame, thus in an absolute way.

THE GENERAL RELATIVITY CONSTANTS G *AND* c *CONSIDERED SIMULTANEOUSLY*

General relativity is often presented as an "extension" of special relativity: Einstein may have been

* *La revolution des quanta* [The quantum revolution], "*Questions de Science*" Series, Paris, Hachette, 1989.

inspired by his 1905 theory to expand on the critique of Newton. The chronology and the vocabulary—as well as some texts—strongly suggest that general relativity emerged from the even more profound refutation of the idea of "immediate action at a distance." Actually, after having attempted to adopt this approach initially, Einstein finally took another one.

His thinking appears to have been guided by another preoccupation that was obviously present in his preliminary work: the desire for simplicity. It should be possible, he thought, to formulate laws for all interactions without having to make mention of the reference frame being used, that is, the classical description of a coordinate system having an origin and axes that allow the location of a point in space to be described. In contemporary terms, we would say that he thought an *invariance* due to a general change of reference frame coordinates exists. Now, to determine such an invariance, he had to postulate a theory of equivalence between the *inertial mass* determined by measuring the acceleration produced on it by a given force and the *gravitational mass* or "weight" that is involved in the law of gravitation. This leads to the conclusion that acceleration due to gravity is independent of mass, and therefore that the constant G is independent of the mass. To complete his program, Einstein had to assume that the same gravitation constant is involved regardless of the mass. Thus, the status of

Newton's constant was modified, giving it greater physical significance.

Einstein was so confident in his general conception of physics, according to which the laws of nature must be as simple as possible, that he wondered whether they could be formulated in such a way that they would be completely independent of any reference frame. This was tantamount to taking yet another step—this time an extremely audacious one—on the path that had led him to the special theory of relativity. Within the framework of this theory, in fact, the laws had been formulated independently of any *inertial reference frame*, that is, independently of any reference frame involving uniform rectilinear motion.

Einstein proposed that we generalize and assume that there is an equivalence between inertial mass and gravitational mass. We can then engage in the following thought experiment (known as the "Einstein elevator"* experiment): a laboratory is enclosed in a box that is moving through space, and a constant force, perpendicular to one side of the box, accelerates its motion. If you found yourself inside this laboratory, the box, accelerated by the application of the force, would appear to you to be a uniformly accelerated reference frame, and you would have the impression that you were being at-

* See Jean-Claude Boudenot, *Électromagnétisme et gravitation relativistes* [Relativistic electromagnetism and gravitation], Paris, Ellipses, 1989.

tracted by the side of the box perpendicular and opposite to the force. However, nothing would allow you to say that you were not falling toward a star that attracts you with an acceleration equal and opposite to the acceleration imparted to the box. Here we must introduce the notion of "field." It entered physics with the work of Michael Faraday (1791–1867) and James Clerk Maxwell who introduced the concept of the "electromagnetic field" to overcome the difficulty encountered in explaining the forces assumed to be acting between electric charges in motion. Hendrik Antoon Lorentz (1853–1928) was the first to attempt the synthesis of Newtonian mechanics and Maxwell's field theory. Einstein succeeded in doing so by calling into question the foundations of classical mechanics: force is replaced by the field described by differential equations. The special theory of relativity allows interactions only in the form of fields. It relies on a field theory of gravitation. A gravitational field imparts the same acceleration to every body, regardless of its mass. The effect of any gravitational field can thus be replaced by a change in reference frame having a uniform acceleration. Conversely, any change in reference frame of this type can be replaced by a gravitational field. It thus appears possible to formulate a theory of gravitation that is invariant under a change in coordinates involving an acceleration. This is the heart of the "general" relativity theory.

We will now resort to a new "thought experiment" to explore the immense implications of this theory. Imagine a laboratory in free fall toward an extremely massive star. Let us set up a very simple experiment in this laboratory by causing a laser beam to propagate in it horizontally. We therefore have a situation where all the equipment in the laboratory is in a state of weightlessness because gravitation exactly compensates for inertia. No force can any longer be exerted. The theory of special relativity thus applies in a reference frame linked to the laboratory; in it, light propagates in a straight line. But such is not the case, however, in a reference frame linked to the star: between the instant that the photons depart and the instant they arrive, the laboratory has fallen. The light thus followed a parabolic path. We can then say that "the light fell," which did not surprise Einstein once he had determined that light carries energy and that energy is equivalent to mass. Since the inertial mass is equivalent to the gravitational mass, the energy of light undergoes gravitational attraction.

However, Einstein adds, since the speed of light is a universal constant, light must always propagate at the same speed regardless of the reference frame. Therefore, if its trajectory is curved, as we assumed in the thought experiment we have conjured up, then the speed of light cannot be invariant! Einstein resolved the difficulty as follows: No, he said, we must hold to the invariance of the speed of light.

What changes with altitude is not the speed of light, but the measure of space and the measure of time (or the measure of space-time): the length of a ruler and the operation of a clock depend on the altitude in a way that the speed of light does not!

The reasoning was as clear as it was audacious: if gravitation is equivalent to inertial forces, then we can ignore it—and also the star—and consider that the only effect of the presence of the star is constituted by a modification of the measure of space-time. Hence, the geometric formulation that Einstein proposed for gravitation: he equated it with the curvature of space-time.

It was then that the problem of the absence of instantaneous action at a distance resurfaced. The curvature of space-time, which is equivalent to gravitation, is *variable*: a gravitational field can only be *locally* replaced by inertial forces. There is no reason to extrapolate from here to everywhere, because there is incompatibility between now and everywhere. Interactions, which are forces exerted by means of fields, physically and actually take time to be propagated. A famous illustration of this fact was given by Ernst Mach. If you twirl around, your arms will extend, and if you also rotate under a starry sky, the sky will also appear to rotate. You would then be tempted to say that because the stars rotate, they are attracting your arms. You would be wrong, because what you do locally does not actually make the stars rotate!

Einstein used this illustration to establish that it is *possible always and everywhere to compensate for gravitation by changing the reference frame, but only locally.*

Let us continue to follow Einstein on the path he took to formulate the laws of physics in a totally invariant manner by a general change in coordinates. We must not fail to emphasize what may have come as a real surprise even to Einstein. His constant and stated ambition was, as we have noted, to make the observation conditions totally and absolutely abstract. He achieved this with the general theory of relativity and, as a bonus, he discovered the relativistic theory of universal gravitation! An approximation of this theory, which takes into account the two universal constants G and c, would consist, for example, in ignoring G to find special relativity; another approximation would consist in making c tend to infinity to obtain the Newtonian theory of gravitation. General relativity can be characterized as the theory in which G is not zero and c is not infinite.

We return to the "falling light" that I referred to in the previous thought experiment. The most extraordinary thing is that this was verified experimentally in 1919 during an eclipse of the Sun. Stars which, according to classical calculations, should have been obscured by the Sun, were in fact visible!

This observation can be explained by the fact that the light coming from these stars is deflected by the gravitational field of the Sun.

Since then, tests of general relativity have been developed. For example, it can be shown that time does not pass in the same way on the first floor of a building as it does on the top floor, because it depends on the gravitational field. Of course, this is an extremely subtle effect—it is evident only in the fifteenth decimal place!—but this effect is real, detectable, and measurable. It would obviously be much more sensitive (by about a factor of two) if we could compare the passage of time on the surface of the Earth and on the surface of a neutron star!

Thus, the theory of general relativity tells us that time depends on the gravitational field. Imagine an extremely massive yet very small star producing an extremely intense gravitational field. Suppose that this intensity is such that the speed of light is less than the escape velocity; trapped, the light cannot escape; the star will be what is known as a "black hole." A photon emitted in our direction by the black hole "falls back" to the center of the black hole, which appears as a singularity in space-time, that is, a point where the equations of general relativity can no longer be applied. There is thus a surface of space-time that we can designate as the "horizon" of the black hole, on which the speed of light seems to disappear (because the photons retrace their path). Since, once again, the speed of light must be treated as invariant, this speed must not vary, much less disappear, and we must assume that, on the horizon of a black hole, time, which is relative to our reference frame, stops or ceases to pass! Since the general theory of relativity does not forbid the existence of black holes, it permits the existence of such *horizons where time stops.*

Einstein's theory can be applied to the Universe freed of its own gravitation. Einstein himself tried it; since he wanted a stable Universe that his equations did not allow, he added to his model an *ad hoc* "cosmological constant" to allow him to stabilize his Universe. Later on, it was observed that the Universe is expanding, and that Einstein's theory could be applied to it without using the "cosmological constant," which he then described as the "biggest mistake" of his life. This is the way that the cosmological model of the "Big Bang" was born, which today is the "standard model" of cosmology. The most interesting aspect of this model is that it has a singularity that appears very similar to the one discovered in the center of a black hole. The only difference is that, in the case of a black hole, it is found in the future of the horizon, whereas for the Big Bang it is found in the past of the horizon.

Nevertheless, there is always a horizon between us and the singularity. The Universe is expanding overall, and the furthest galaxies are moving away from us the fastest. At a certain distance, their escape velocity is equal to the speed of light. This means that light emanating from these galaxies cannot ever reach us: a space-time surface will indicate where this speed will appear to us as having disappeared, and thus where time will stop.

This clearly demonstrates what I proposed: The two universal constants G and c mark absolute limits: the same now cannot ever be defined everywhere.

II

THE QUANTUM CONSTANTS *h* AND *k*

The constants G and c allow us to consider our relationship to the infinitely large, that is, to the entire Universe. But this Universe appears to us as a series of interlocking structures whose sizes extend over more than 40 orders of magnitude. If we now turn to the infinitely small, the obvious question that arises is how the limits of this interlocking can be determined. Is there a limit to the divisibility of matter? This ancient question continues to stimulate contemporary physics thinking.

STATISTICAL THERMODYNAMICS AND THE BOLTZMANN CONSTANT k

We know that one of the greatest advances of human thought is the "atomic hypothesis" which, since Leucippus and Democritus in the 5th century B.C., has postulated the existence of such a limit. The great works of the Austrian physicist Ludwig Boltzmann (1844–1906) fit into the framework of this hypothesis and extend its consequences to the extreme. A new constant, called the Boltzmann constant, resulted from the efforts of researchers to apply the laws of mechanics to a very large number of atoms. In this way it was discovered that the enormity of the number of "degrees of freedom" governing the system under consideration, far from being an obstacle to the theory, as one would suppose, opens the way to a remarkable simplification. In general, a degree of freedom is any parameter in a given system that can vary independently of the others. It can involve the position of a point on an axis, of a coordinate in space, of an angle defining an orientation, etc. A system with zero degrees of freedom would be composed of an isolated motionless point. This is an "ideal" case because it is not possible for such an object to actually exist. A system with one degree of freedom is a discrete object moving along a line. Now imagine a system composed of a gas enclosed in a box. If we accept the atomic hypothesis, we would say that this gas is composed of an enormous number of molecules. The position of each molecule, described by three coordinates, represents three degrees of freedom. The typical number of molecules contained in a macroscopic box was determined by the Italian chemist and physicist Amedeo Avogadro (1776–1856): for standard volume, pressure, and temperature conditions, it is 6.022×10^{23}. We can thus determine the order of magnitude of the number of degrees of freedom in such a system: it is enormous. But this system, which has such a large number of degrees of freedom, nevertheless presents qualitative macroscopic properties that can be understood and described.

Boltzmann's brilliant idea was to use rational mechanics approaches to account for these properties. He had used the idea of "phase space"—an abstract space where a system is represented by a point. This involves a number of dimensions: in the phase space there are as many dimensions as there are degrees of freedom in the system. Actually, there are twice as many, because phase space is composed of not only their degrees of freedom but also their speeds, that is, the derivatives of these degrees with respect to time. The idea implemented by "rational mechanics," that is, by mechanics mathematized, as we mentioned above, by Lagrange and Hamilton, was to replace the problems posed by the description of the evolution of a system by the study of the trajectory of a point representing this system in this abstract phase space.

Thus, an extremely complex system can be represented by a single point in a space that can have an enormous number of dimensions. In mechanics, the equations of motion for this representative point are established in this way. This motion is determined by these equations and by the initial conditions or the conditions at the limits.

Let us now imagine the case we considered earlier: a gas enclosed in a box, which is thus a system subjected to an extremely large number of degrees of freedom. It is then necessary to have an extremely large number of equations to account for its evolution, but they are fundamentally of the same type as those that govern the evolution of any other mechanical system. All the difficulty arises from the fact that it is extremely difficult to specify the initial conditions. This problem can be resolved only by considering the regions in phase space, then by using statistical, probabilistic reasoning. Thus we define what is known as "phase density," that is, the probability that the representative point of the system will be found at a certain point in phase space with a certain precision in a certain infinitesimal "volume" (since phase space has more than three dimensions, it is a "hypervolume"). We thus obtain all the information that is humanly possible to acquire with regard to the system in question. The "phase density" therefore represents the probability distribution in phase space.

From Boltzmann, and then Willard Gibbs (1839–1903) and Albert Einstein, came the idea that only a small number of physical quantities, such as total energy, temperature, or pressure, is needed to completely characterize the macroscopic properties of a system, whereas an unimaginably large number of microscopic configurations, known as "complexions," gives rise to the same macroscopic state.

Let us acknowledge that the number of degrees of freedom is enormous, and that we are dealing with molecules that move in all directions; the total kinetic energy of the system will be zero, but it will retain internal energy, namely, the heat related to the temperature. Because of the very large number of degrees of freedom, averages can be determined. What physical quantities will be adapted to the description of such a system? Energy will be at the top of the list, because it is a concept that is just as operative at the macroscopic level as at the microscopic. Then we will define a physical quantity that allows us to consider the number of complexions resulting in the same macroscopic state. This quantity, which is well known today, is called *entropy.* It involves a physical quantity of a new type because it cannot be expressed by using fundamental physical variables (length, time, and mass). Its dimensional content is actually defined by energy divided by temperature; it is a physical quantity that has meaning only on a macroscopic level. For this reason, many physicists refuse to consider it to be a fundamental quantity. This reluctance reflects the debates that the notion of entropy continues to cause as a consequence of its subjective content. Since the dimensional content of Boltzmann's constant k is exactly the same as that of entropy (Boltzmann's constant appears in the equation $E = kT$, where E is energy and T is temperature, its value is 1.380×10^{-23} joule per kelvin), we can understand the reluctance of those who consider k to be a universal constant just as fundamental as the others.

In Boltzmann's time, the second law of thermodynamics was known as the "Carnot-Clausius Law." This law appeared in science as the expression of a practical impossibility: that of constructing a heat engine with an efficiency greater than or equal to 100%. Sadi Carnot actually discovered that heat cannot be transformed into energy without a loss of energy. His *Réflexions sur la puissance motrice du feu et sur les machines propres à développer cette puissance* [Reflections on the Motive Power of Fire and on Machines Fitted to Develop that Power] (1824) hypothesized that every mechanical process is accompanied by a loss of energy in the form of heat. This principle destroys the temporal symmetry accepted by classical mechanics; it orients time. Time now has a direction, it travels in the direction of the increase in entropy. What Boltzmann succeeded in doing was to formalize and express in quantitative and precise terms that this inevitable increase in entropy is due to the information lost when a system is described by using macroscopic physical quantities. This is explained by the fact that, when the number of degrees of freedom is large, there is an enormous number of complexions that give rise to the same macroscopic property and that cannot be determined. Thus, entropy appears as lost information.

However, this "modern" language must not deceive us: the informational interpretation of entropy was formulated long after Boltzmann, within the framework of a theory which at the outset had nothing to do with thermodynamics, namely, the information theory developed in 1948 by Claude E. Shannon, an engineer at Bell Telephone. According to this famous theory, the information content of a particular message can be defined by means of "statistical entropy." If we wish to give this information the dimensional content of entropy defined by thermodynamics, we can say that statistical entropy corresponds to a particular probability. But probability is a dimensionless number; it is then necessary to use an additional constant. This constant is the "Boltzmann constant" *k*. In other words, if the Boltzmann constant is used in the statement of the law that defines statistical entropy by using probability, then statistical entropy is identical to thermodynamic entropy. I must emphasize that it is essential to avoid any anachronism about the time frame of these events; even if Boltzmann actually had the idea for statistical entropy, he could not have had access to information theory. It is obvious that Boltzmann's work was awaiting a reformation which now allows us, as we shall see, to interpret his constant as a *quantum of information*.

For now we need only accept that the atomic hypothesis can be made compatible with thermodynamics as it was developed by Carnot and Clausius with rational mechanics, by using the concept of entropy which can be given a statistical interpretation. As Newton's constant in its domain once did, Boltzmann's constant appears at first glance to have only a utilitarian content: its only function is to provide this compatibility between the atomic hypothesis and thermodynamics.

PLANCK'S CONSTANT h, *THE QUANTUM OF ACTION*

It is often said that Planck's constant resulted in the appearance of the discontinuity in matter that suddenly and forever disconcerted physicists. Actually, the discontinuity discovered by the German physicist does not affect matter but rather interactions and forces. The biggest surprise is that, although the atomic hypothesis still stimulated many debates at the beginning of this century, it is nothing more than the hypothesis of the discontinuity of matter, having no radically novel character; it was already subsumed under thermodynamics and, as we have recalled, it had already guided many of the most eminent physicists and allowed them to obtain remarkable results.

But a discontinuity residing in what are today called interactions, that is, in forces, was much more difficult to accept and caused a veritable crisis in physics thinking! What idea were we to have of what a force is? An idea, inherited from Newton, according to which we can always make a force tend to zero. The continuous character of the force seemed to be a part of its own definition. For example, this is how the well-known gravitational force appeared, as well as the electromagnetic force described in 1865 by the equations of James Clerk Maxwell. Physicists, among them Max Planck, found themselves forced to re-examine questions they believed had been resolved by Newton, Leibniz, and Descartes' disciples in the 17th century!

To shed light on the difficulty of the question—and on the intellectual feat that allowed it to be overcome—we need to speak in the language used by physicists today. Let us say that we found it necessary to introduce discontinuity into an "interaction." This is not a concept, but rather what I would call a "category" that designates "*in vacuo*" everything involved in the formation of a structure—its evolution, its stability, or its disappearance.

This notion is used in such a broad sense that if it is said that the Universe is a series of interlocking structures, then it can also be said that everything is the result of interactions. Thus, electromagnetic radiation belongs to the category of interactions. And it is the problem of the thermodynamics of radiation that

Max Planck addressed. We know that he suddenly encountered a problem, a surprising anomaly: "black-body radiation." According to classi "cal physics, the emission and absorption of light by matter occurs in an absolutely continuous fashion. The amount of luminous energy must thus flow continuously like a fluid. However, Planck observed that when a closed box, capable of absorbing equally all radiation regardless of frequency, is brought to a certain temperature, it emits radiation *discontinuously*, by "discrete" values, by "quanta." A particular interaction, in this case, radiation, having a wavelike nature, proved to have properties previously reserved for thermodynamic phenomena! This contradiction could be resolved only by imagining that electromagnetic energy is carried by "particles," so that the energy is proportional to the frequency of the radiation. This turned out to be the theoretical role for the constant formulated by Max Planck: to relate energy and frequency, according to the famous equation $E = h\nu$, where E represents energy, ν the frequency and *h* "Planck's constant," which equals 6.622×10^{-34} joule-second.

This was such a radical revolution in physics thinking that Planck was at first overwhelmed by the consequences, and it took all the audacity of a young Albert Einstein to interpret *h* by introducing discontinuity into interactions. From 1900 to 1905, he attempted to treat electromagnetic radiation like a *material* whose divisibility had limits. He treated the difficulty he had in mechanically interpreting the wave theory of electromagnetism by proposing a statistical calculation for radiation. But how could this be applied to waves? This seemed to be totally impossible unless the existence of "particles" of light (photons) was assumed.

It was here that experimentation overtook speculation to confirm "logical" conclusions. Because, at the same time as the confrontation between theory and blackbody radiation arose, so did the question of what is called the "photoelectric effect," which appears as a frequency threshold. A photoelectric body emits current when it is irradiated by luminous radiation. But below a certain frequency, even if the intensity of the radiation is greatly increased, no current is obtained. Above this frequency—this threshold frequency—the current is maintained even when the intensity of the radiation is decreased. In 1905, Einstein was the first to propose the idea that the energy of electromagnetic radiation is carried by these particles which he called "photons" and for which the equation $E = h\nu$ relates frequency to energy. The frequency threshold then corresponds to an energy threshold: for the photons to be able to produce current by removing electrons, they must have sufficient energy. To this "effect" we must add another—no less famous—one: the "Compton effect," named after the American physicist Arthur Holly Compton (1892–1962), who demonstrated it twenty years later. This effect shows that an electron recoils when struck by a photon, and that this recoil obeys the laws of mechanics, which proves that a collision between an electron and a real particle is involved. The existence of the photon became indisputable, since it could somehow be observed, even if the observation was not direct.

At the beginning of this century, in a matter of a few exciting years, an amazing fact was discovered, namely, that an exchange occurs with every interaction and, moreover, that there is a minimum exchange below which there is no further interaction. In other words, the idea being developed was that it is not possible for two structures or two systems to interact without exchanging something; that there is a smaller "something," a *quantum of interaction* which must be exchanged for an interaction to occur.

Let us imagine that we have a device that allows us to "see"—observe and experiment with—atoms. This is no longer what Einstein called a simple "thought experiment"; today we have at our disposal "detectors," along the lines of, for example, a Geiger counter, named after the German physicist Hans Geiger who invented it in 1912. An atom detector would be a device with at least one of its parts in a state of hyperunstable equilibrium, so unstable that the perturbation of a single one of the atoms would suffice to destabilize it. The detector would then be capable of transforming this "microscopic" perturbation into a macroscopic signal. An elegant example is a "Wilson chamber," named for its inventor, the Scottish physicist Charles Thomson Rees Wilson (1869–1959). The device is extremely ingenious and admirable in its simplicity. In a vapor that is close to condensation, electrically charged particles in motion ionize atoms along their entire pathway, to such an extent that liquid begins to condense near these ionized atoms, with small droplets forming and tracing the path of the particles in a display of great beauty!

Let us suppose that a device of this type allows an atom to be observed and thus has a macroscopic effect in response to a microscopic perturbation. How is this response to be interpreted? According to atomic theory, this instrument is itself composed of atoms. For it to provide a macroscopic response, at least one of these atoms must receive a signal from the emitting atom—the one that will be designated as the "object." Assuming that the detector is sufficiently efficient, precise, and sensitive, the amplification process of the signal leading to its macroscopic "translation" occurs without further perturbing the emitting atom. The fact remains that a signal had to be emitted, received, and amplified to become macroscopic. However, this is not sufficient reason to believe that the signal was, on the atomic scale, marginal or insignificant. On the contrary, the emitting atom, after having emitted its signal, will not be found in the same state it was in before the emission occurred. In other words, to obtain a result that can be measured, it is necessary to establish a minimal coupling between the object and the device, a coupling that is not marginal on the atomic scale and thus inevitably and irremediably perturbs the object. This perturbation cannot be made to approach zero if we are still to obtain a result we can measure! The physicist can thus cause the quantum of interaction to appear in a practical act that is essential to scientific activity: the experimental act. We can go even further and assert that an experiment is always just an interaction, and that there is always a minimum "coupling" between the object and the device—something which the "classic" ideal of science has too often forgotten or deliberately "neglected."

Can we give a quantitative interpretation to such an idea? We can, if we refer to a concept that has proven extremely useful with systems having a small number of degrees of freedom in mechanics as well as in thermodynamics: the concept of "action." Action is defined as the product of energy and time; and, in fact, all the laws of classical mechanics can be explained by the principle known as "least action", which was proposed for the first time in the field of optics by the French mathematician Pierre Louis Moreau de Maupertuis (1698–1759). "In any change that occurs, the quantity of action necessary for each change is the smallest possible," wrote Maupertuis in 1746, after having defined action two years earlier as "the product of mass, space traveled, and speed."

To understand the importance of this principle in contemporary terms, we must refer to the notion of "phase space": the trajectory of a representative point of a system in such a space has the property of minimizing a quantity which has the dimensional content of an action, called the action integral. It is possible to demonstrate that this principle of least action is rigorously equivalent to the Lagrange and Hamilton equations of rational mechanics. Note that action is different from power. Power is defined as energy divided by time; action is energy multiplied by time. The reader can verify in everyday experience that when a rather heavy load is carried over a certain distance, the product of mass and speed and length is in fact the same as energy multiplied by time. Maupertuis readily took to using terminology borrowed from economics, and so can we still today by saying that action is defined by work time, which is to say energy, or work, multiplied by time.

We now return to the coupling between our "detector" and the microscopic object it allows us to observe. We can measure the perturbation of the object with an energy. Let us call the energy provided to the emitting atom that is necessary to obtain the result of the measurement ΔE. ΔE can be positive or negative. Can we make ΔE approach zero while continuing to obtain a result? Undoubtedly, but on the condition that the experiment is carried out over an infinite time! The action ΔA, which is the product of the absolute value of the energy perturbation ΔE and the duration ΔT of the experiment, cannot be made arbitrarily small. Assume that the work time necessary to obtain the response cannot be reduced to zero. We can explain this impossibility by saying that there must be an action that represents a lower limit of all action in all of the Universe, somewhat similar to the way that the absence of instantaneous interaction is a manifestation of the existence of an upper limit for all speeds (the speed of light). This is what is known as the "quantum of action," equal to Planck's constant h.* This quantum is written in Heisenberg's first inequality ($\Delta A = \Delta E \Delta T \geq h$) which manifests the existence of an inevitable coupling between the object and the measuring device.

This formulation was suggested to me by the work of the Soviet physicist Lev Landau as well as the reflections of Niels Bohr in the many texts he devoted to his great contributions to quantum theory. Together with Michel Spiro, I developed some of these ideas in *La Matière-Espace-Temps* [Matter-space-time] and with Jean Pierre Baton in *L'Horizon des particules* [The horizon of particles]. It is impor-

* Since frequency is inversely proportional to time, the relation $E = h\nu$ implies that h has the dimensional content of action.

tant to emphasize the epistemological consequences of this formulation, because there is, in fact, a quantum of interaction present in all experimental interactions. We can therefore no longer treat the observation conditions abstractly. One of the great dogmas of classical physics thus collapses; according to this dogma, still supported by some physicists, there is no "subject" of scientific knowledge.

Actually, classical physicists opposed this dogma even in their own practice. And Einstein, who was relentlessly blamed for it, had never even bothered to think about it. He had certainly openly and repeatedly declared his ambition to arrive at an entirely intrinsic description of reality; but, as we have seen, he was practically required to renounce this goal after he determined in his theory of general relativity that if a reference frame can be found that exactly compensates for gravitation, then this can only occur locally.

In any case, in physics today, we can no longer pretend to ignore the observation conditions. This, in fact, was the opinion of Bohr, who said that with the new physics, we can no longer pretend to describe reality itself directly; physicists describe what he proposed to call a *phenomenon* as an element or a moment of reality, placed under observation conditions that are as well-defined as possible. This carefully considered and expressed position does not in any way deny the existence of a reality independent of observations, as many impatient philosophers wanted to understand this.

Bohr's hypothesis, to which I fully adhere, thus leaves the methodology of quantum mechanics entirely open. If we now consider the observation conditions, we can in fact vary these conditions in a controlled manner in order to refine the representation of reality. The fact remains that a considerable amount of work was required for the observation conditions to be taken into account within the very formalism of physics; debates and controversy still continue almost a hundred years after Planck's discovery.

The difficulty involved in taking the observation conditions into account is that the flexibility of physics concepts must not be impeded. It is obvious that if a concept purported to describe, for example, the behavior of an elementary particle were to contain *explicit* references to the experimental device used in the observation, the concept would be totally unusable. The extraordinary efficiency of quantum formalism resides in the fact that the reference to observation conditions is maintained *implicitly* and that it does not impede in any way the use of the concepts.

Quantum concepts, as we said, are relative to "phenomena," if by this word we mean the reality "as it exists." However, the observation conditions depend on macroscopic instruments, which contain an enormous number of atoms and particles of the same size as the object being described. Consequently, these "observation conditions" can best be determined *statistically.* The really brilliant idea brought to bear on this problem consists of using *probabilistic predictability* instead of the so-called "deterministic" probability of classical physics to describe microphysical reality.

This radical innovation has led to and continues to result in misunderstandings, or even fierce resistance. Nevertheless, this innovation only emphasizes the fact that *deterministic concepts would not be credible* if the observation conditions, which the existence of the quantum of action constrains us to take into account, can best be determined statistically. In any case, as Léon Rosenfeld* said, recourse to probability does not in any way imply any renunciation of scientific objectivity: "Probability does not mean randomly without rules, but precisely the contrary: there are rules of randomness. A statistical law is first and foremost a law, the expression of regularity, and an instrument for prediction."

THE BOLTZMANN CONSTANT AS A QUANTUM OF INFORMATION

In my opinion, the only way we can overcome these difficulties is to use Boltzmann's constant *k* which is

*A. George, *Louis de Broglie, physicien et penseur* [Louis de Broglie, Physicist and Thinker], Paris, Albin Michel, 1953, p. 58.

necessarily, albeit implicitly, included in the determination of the observation conditions. Here, the French physicist Léon Brillouin (1899–1969), who became a professor at Harvard and then director of research at IBM, paved the way by proposing an interpretation of the Boltzmann constant as a "quantum of information," which he was able to do only as a result of developments in quantum theory. His approach consisted in exorcising "Maxwell's demon." In concrete terms, he used the allegory that claims to show that the second law of thermodynamics has no objective content and only expresses a limitation, which is not a law, of our means of knowledge. Maxwell imagined a closed box containing a gas at thermal equilibrium. This box is divided by a wall containing a small trap door. Suppose, he wrote, that there is a hypothetical demon moving around in this box who knows how to open and close the trap door without expending any energy. Each time a fast-moving molecule is seen, it is allowed to pass into the left-hand side of the box; when a slow-moving molecule comes along, it is allowed to pass into the right-hand side of the box. After a certain amount of time, the demon will have thus created a temperature difference between the two sides of the box, and can then extract mechanical energy from it.

According to Brillouin, if this supposed demon does nothing, it cannot see the molecules. What Maxwell did not know was what blackbody radiation theory has since established: in the box, the molecules cannot be seen because the radiation is blinding. However, added Brillouin, give Maxwell the benefit of his historically unavoidable ignorance and attempt to extend his idea by allowing his demon to use a flashlight to see the molecules one at a time. Brillouin showed that the demon would then confront the quantum of action: it cannot see a molecule without modifying its energy and thus without acting on the entire system. If it is true that the demon can obtain information about a molecule, it will increase the entropy of the rest of the molecules. Thus, entropy is equivalent to lost information; we can also say that information consists of "negentropy" (remember that entropy and information have the same dimensional content). Hence, by attempting to decrease the entropy by measuring the position of a molecule, the demon, caught in a trap, creates entropy! Even when the information is equal to k (or, as Brillouin showed in a more precise analysis, equal to k multiplied by the logarithm of 2), the demon causes more entropy to be created than information gained! Although Brillouin never explicitly used the term, he interpreted Boltzmann's constant as a *quantum of information.* This involves the smallest quantity of information associated with a microscopic degree of freedom, in other words, a quantity below which there is no information, a quantity of information that is irremediably lost and which becomes a quantum of entropy if this degree of freedom is not determined.

How does the existence of such a quantum of entropy give a fundamental character to the second law of thermodynamics? In my opinion, it is by a subtle relationship that is established between the quantum of entropy and the quantum of action. The quantum of action appears as the *cost of the quantum of information*, which, moreover, is compatible with the "economics" interpretation that we gave above for the concept of action. Recall the reasoning presented above to introduce the quantum of action: for the detector to give a macroscopic response, at least one of the atoms must have received a certain signal from the object. The information content of this signal is a quantum of information. The cost of this quantum of information is the quantum of action. The second law of thermodynamics, which strictly precludes 100% efficient heat engines, frictionless motors, and perpetual motion machines, reflects the existence of this quantum of cost. In fact, according to atomic theory, every macroscopic system has a very large number of degrees of freedom, and if all friction were eliminated and assuming that the unimaginably large number of complexions could all be distinguished from one another, it would have an almost infinite cost.

However, Brillouin noted that in exorcising Maxwell's demon, if he had to use the quantum of action, Planck's constant would not explicitly appear in the expression of the quantum of information.

Thus we see the subtle dialectic of h and of k. The quantum theory for systems with a small number of degrees of freedom consists in considering the universal constant h. This assumes taking into account the observation condition which applies with the possible use of macroscopic devices. However, the existence of the quantum of entropy implies that these observation conditions can, in principle, best be considered statistically. This difficulty can be bypassed by recognizing that the predictability of quantum concepts, even when they relate to systems with a small number of degrees of freedom, can only be probabilistic. To establish quantum formalism, it is therefore necessary to use k, even though this constant is present only *implicitly.* On the other hand, we have seen that to give fundamental content to the second law of thermodynamics, we need quantum of action, even if Planck's constant disappears from the formalism; quantum theory only *implicitly* underlies rigorous formalization of statistical thermodynamics, but it is absolutely necessary for it. In fact, the third law of thermodynamics, which states that a system at a temperature of absolute zero has an entropy equal to zero, is strictly of quantum origin.

With Jean Pierre Baton, I stated the hypothesis, in *L'Horizon des particules* [The horizon of particles], that the relationship between the two universal constants h and k can be expressed by using a new constant equal to the ratio of the two, which we propose to call constant b in honor of Brillouin: $b = h/k$. This constant has the dimensional content of temperature multiplied by time. Such a physical quantity expresses the cost of information in terms of action. However, if action and information each has a quantum, then b is a *quantum of cost.* The physics in which such a quantum of cost existed would of necessity be probabilistic: in fact, infinite precision necessary for a deterministic predictability would be impossible because it would be infinitely costly. This characterization can be applied, as we just saw, to statistical thermodynamics and quantum mechanics, which consider, even if each only implicitly, the constants h and k, and of course their ratio. However, this characterization can also be applied to the description of chaotic processes that have been discovered in the dynamics of nonquantum and nonthermodynamic systems, because they are macroscopic and have only a small number of degrees of freedom, like coupled and maintained pendula, etc. The evolution of these systems is chaotic because it is *sensitive to initial conditions*. However close they may be, even initial conditions that are not strictly identical lead to evolutions that inevitably diverge. The evolution can therefore be predicted during an infinite duration only at the price of an infinite precision in the determination of the initial conditions. However, quantum of cost renders this infinite precision impossible. It is interesting to note that such chaotic physics is obtained at the limit where h and k approach zero (a limit that is neither quantic nor thermodynamic) but with a finite quotient b.

Thus, we see how the significance of the two constants h and k emerges. They are both evidence of a limitation of the principle of human knowledge: *all knowledge has a cost.* The time during which it is possible to predict the evolution of a system necessarily proves limited. In this way, another "demon" is exorcised—the demon of Laplace, who, knowing at a given moment the position and speed of all the particles in the Universe, was supposed to have been capable of predicting its *entire* subsequent evolution. Contemporary physicists have dethroned this sovereign "intelligence" suggested at the beginning of the "*Essai philosophique sur les probabilités*" [Philosophical essay on probability], where it is believed that the first statement of the "deterministic" vision of science appeared. The universal constants h and k delimit a temporal horizon; they express the incompatibility of *here* and *forever.*

The presentation we have made up to this point for the four universal constants is essentially negative. These constants express the inherent limitations of human knowledge, that is, the impossibility of extrapolating from "here and now" to "everywhere and forever." Does this mean that all science will now become impossible? The liveliness of the debates that have animated physics since the beginning of the 20th century demonstrates that the question has been taken seriously, and that it is not self-evident that the response to it should be negative. The quantum revolution has allowed a tremendous reversal, which we will now describe: as a result of the quantum revolution, the inherent limitations of knowledge have become anchor points for new scientific advances.

PROBABILITY AMPLITUDES AND WAVE-PARTICLE COMPLEMENTARITY

We have already discussed the principle of this reversal: because we can no longer ignore observation conditions, it is appropriate to include them in the content of these concepts. The entire problem will be that this acknowledgment should not interfere with the flexibility of the concepts.

It seems to me possible, without in any way betraying the thinking of Niels Bohr (1885–1962), to reformulate the requirement of taking into account the observation conditions that he included in the notion of *phenomenon*. I propose to use the notion of *horizon* which shares with Bohr's phenomenon the property of uniting the "subjective" and the "objective." A horizon is objective, because if there were no world, obviously there would be no horizon; but it is also subjective, because it depends on the position of the observer for whom it is the horizon. The horizon line separates the world into two parts: the part on the side of the observer, which we could call the "proper world of the observer," or the domain of *actuality*; and the part that is beyond this proper world, or the domain of *potentiality*. The horizon line appears fictitious, virtual, unreal; it exists only relative to the observer. It is inaccessible, because it always retreats; it is mobile, and its motion parallels the motion of the observer. *It can move at the speed of light or at the price of a simple quantum of action.* Nevertheless, we know one thing for certain with regard to the horizon line: we know that *it is definitely in the world that we draw it.* It is precisely on this certainty that the fundamental reversal brought about by the quantum theory rests: the real world is now thought of as *the place of all possible horizon lines*; quantum theory thus proves to be the theory of the motion of horizon lines. I label the quantum concepts as "horizontal" to indicate that they describe not the "real world" in the sense that this expression is ordinarily understood in accordance with a long philosophical tradition which "makes" or "substantializes" the real, but its horizon lines.

We have seen that quantum theory had to renounce deterministic predictability in favor of probabilistic predictability. But the acknowledgment of the quantum of action could not be demonstrated through the use of classic probability theory; it was necessary to construct a horizontal probabilistic concept.. This is how the concept of *probability amplitude* arose. It is a "complex number" that by definition has a real part and an imaginary part, or, alternatively, it can be defined by a modulus and a phase, where the square of the modulus (or the sum of the squares of the real part and the imaginary part) is a probability (that is, a positive real number less than 1). In other words, a probability amplitude is the "complex square root" of a probability. It thus unifies an *actuality* and a *potentiality*: the actual pole is located in the modulus, directly linked to the probability, which can be measured by performing repeatable experiments; and the potential pole is in the phase, which is not measurable but allows a given probability to be associated with an entire class of potentially equivalent amplitudes which differ only in their phases.

The horizon for which probability amplitudes describe the lines is that of *quantum discernibility.* Two states of a system, or two transition pathways, are said to be quantically indistinguishable, if to make a distinction between the two costs at least one quantum of action. The essence of quantum theory formalism consists of a strict use of probability amplitudes. The golden rule that summarizes it can be stated as follows: in the domain of indiscernibility, probability amplitudes are added like complex numbers (algebraic sum of the real and the imaginary parts); in the domain of discernibility, the probabilities are added.

The rule for adding complex numbers for probability amplitudes in the domain of the indiscernible allows the new formalism to be adapted to the wave-particle duality which is one of the most disconcerting characteristics of the quantum universe.

We have seen that, in classical physics, the electromagnetic interaction is described by using the Maxwell equations that govern wave dynamics. But the descriptions of blackbody radiation, the photoelectric effect, and the Compton effect require a *particle* representation of the electromagnetic interaction. The major idea of Niels Bohr and the Copenhagen School was that of the *complementarity* of the wave and the particle points of view: it is the same quantum reality that, depending on the observation conditions (again!), reveals the wave or particle aspects to the observer. This complementarity results in extremely unsettling paradoxes. Consider, for example, Young's classical double-slit experiment* which causes *interference*, a specific

*A mask pierced by two slits is placed between a point light source and a screen.

effect of wave dynamics, to appear. It is well known that the waves of an electromagnetic field can interfere, that is, produce alternating dark and light bands on the screen. But if this interaction requires a particle interpretation, how can there be any compatibility between this interpretation and the effects of interference? If electromagnetic radiation is compared to a photon flux, how can the photons interfere in the same way waves do?! If the electromagnetic wave can be divided into two parts, each of which passes through a slit and then interferes, the photons, being indivisible, must each pass through one slit or the other. They cannot interfere!

Nevertheless, it just so happens that the probability amplitudes can interfere. Quantically, the paths a photon can take through one of the two slits are *indiscernible* paths, in the sense that we just described, because to know through which slit the photon has passed, it is necessary to perform an experiment that would cost at least one quantum of action. Since the two paths are thus indiscernible, the corresponding probability amplitudes must be added as complex numbers to give the total amplitude of the probability of a photon's impact at a certain point on the screen. Since complex numbers can "interfere" constructively or destructively, we can observe on the screen alternating areas of large and small probability of photon impact, that is, areas of light and dark bands.

However, probability amplitudes can also be used to study the discernible paths. An experiment allows us to confirm this. We can, in fact, design and construct devices to detect through which slit the photons pass by counting only the impacts for which we are sure of the slit through which they passed. And we would then discover that the interferences disappear! Such disappearance indicates that, for discernible paths, it is necessary to add the probabilities, without the possibility of interference.

BOSONS AND FERMIONS

Louis de Broglie (1892–1987) went even further on the path of quantum complementarity for particle and wave dynamics. He said that since classical wave phenomena have a particle representation in the quantum domain, why not look for a wave representation in the quantum domain for classical particle phenomena? For this reason he proposed associating an electron flux with a wave packet. The kinematic characteristics of particles, the energy E and the momentum (or quantity of motion) p, are related to the characteristics of waves, the frequency ν and the wavelength λ by the Einstein equation $E = h\nu$ and the de Broglie equation $p = h/(2\pi\lambda)$, where h is Planck's constant. The formalism of the probability amplitudes described above allows us to give a quantitative form to this association. Thus, the reasoning that we outlined for the representation of the Young experiment in terms of photons can be repeated word for word for a "Young experiment with electrons." Such experiments have now been carried out; they clearly validate the generalization of the wave-particle complementarity applied by de Broglie.

We are beginning to see with some clarity how a certain awareness of the limitations that affect the principle of knowledge allows us, without any paradox, to refine the knowledge we obtain based on the real. To think beyond the horizon of discernibility, we have created the concept of probability amplitude. By enlisting this concept, physicists can, without eliminating this horizon, make it move, and thus discover the unsuspected properties of the real world. For example, let us take the particle aspects in interactions or the wave aspects in matter, which enrich and clarify in greater detail their representation of the real world because they can now draw on two available representations—the wave representation and the particle representation—each of which unifies matter and its interactions.

However, if it is true that there is a particle representation for interactions, we have every reason to believe that the interacting particles are not completely identical to the particles of matter. We then look for a criterion that would allow a distinction to be made between a particle associated with an interaction and a particle associated with matter. The formalism of probability amplitudes provides us with such a criterion; we can use it to distinguish "bosons" from "fermions." "Bosons" are interacting particles; they obey the Bose-Einstein statistics. An arbitrary number of them can be found rigorously in the same quantum state. "Fermions," on the other hand, are particles of matter which obey the statistics of Fermi and Dirac. Their behavior is such that only a single one of the same kind can occupy the same quantum state. They are governed by Pauli's "exclusion principle." This principle is based on the classical and even ancient idea—but in a completely new form—that particles of matter are impenetrable. It provides a satisfactory explanation for the fact that matter does not collapse on itself under the effect of gravity. The interaction waves are superimposable. This characteristic is perfectly classical, and it is expressed in quantum theory by the fact that when they are in the same state, an unlimited number of interaction particles can be present. Note that the more bosons in the same state, the greater the probability of the state. It can be said, using social imagery, that if the fermions are "individualistic," the bosons' behavior is "gregarious." Lasers, which are well-known to the general public, use this "gregarious" property to produce extremely coherent light.

However, we still have not understood how the two universal constants h and k can be considered simultaneously and *explicitly.* This is precisely what contemporary physics is doing: considering h and k simultaneously by rationally coordinating them in what is known as "quantum statistics." A distinction can thus be made between the domain and the objectives of a domain of physics concerned with systems having a large number of degrees of freedom and in which quantum effects exist. This research yields results that are spectacular and rich in technological implications, such as low-temperature physics, superfluidity, and superconductivity, to name but a few.

Quantum theory is probably one of the most notable aspects of 20th century theoretical thought. To attempt to understand its content, without dealing too much with the inevitably esoteric technical developments, I have referred to the idea of horizon, which I use neither as a simple metaphor nor as a strictly scientific concept, but rather as a "paradigm." So it appears that, as attested by the three following texts, this paradigm can play a decisive role not only in the philosophy but also in the poetry of the 20th century.

In one of his major works, the Swiss philosopher Ferdinand Gonseth wrote, after having set forth the relationship among intuition, axiomatization, and experimentation in modern geometry: "The foregoing results have a value that transcends the scope of geometry. They concern the whole of knowledge or, as we would prefer to say, the state in which all knowledge appears to us at a given moment: nothing authorizes us to believe that our knowledge, even at its outermost limits, is any more than a *horizon of knowledge*, and that these most recent *realities* that we have conceived are any more than a *horizon of reality.*"*

Maurice Merleau-Ponty, one of the most eminent exponents of *phenomenology*, a major trend in 20th-century philosophy, writes: "It is in the test I am conducting of an exploratory body intended for things and for the world—of a feeling that I sense in my very core and which immediately draws me from quality to space, from space to things, and from things to the horizon of things, that is, to a world that already exists, that my relation with being is bound."**

The book by Michel Collot, *La Poésie moderne et la structure d'horizon*, also deals with these philosophical concerns and has caused me to re-examine my thinking about the horizon. In it I read: "The

---

* Ferdinand Gonseth, *La Géométrie et le problème de l'espace* [Geometry and the problem of space], Neuchâtel, Ed. Le Griffon, 1945/1955, p. 310.

**Maurice Merleau-Ponty, in his report on his work presented to the Collège de France in 1951, cited in *Phénomenologie de la perception* [Phenomenology of perception], Paris, Gallimard, coll. Tel.

reality to which the poem refers is not that of the objective Universe that the sciences attempt to create, but rather that of the world as it is perceived and experienced. Thus, the latter never appears as the *horizon*, that is from the particular viewpoint of a subject, and according to a changing articulation between what is perceived and what is not, between the elaboration of a structure and the opening of a vast expanse of uncertainty."* I have the impression that I could rewrite these lines simply by substituting the thinking of physics for the word "poem."

* Michel Collot, *La Poésie moderne et la structure d'horizon* [Modern poetry and the structure of the horizon], Paris, PUF, 1989, p. 7.

III

THE MARRIAGE OF RELATIVITY AND QUANTA

Ever since relativity and quanta appeared, the problem of the marriage of these two major theories has been presented to physicists as a challenge that will decide the future of their discipline. The fierce debate between Einstein and Bohr that lasted several decades is related to this difficult marriage. Where are we today?

We can say that the challenge has been partially accepted. On the whole, we have actually succeeded in uniting the theory of *special* relativity and quantum theory by simultaneously taking into account h and c. The result of this union is the *standard model of elementary particle physics and fundamental interactions*, which was finally worked out during the 1970s. We know that there are four "fundamental" interactions, namely those which cannot be explained by any other interaction. After having thought that electric and magnetic forces were distinct, we were able to determine that these are just two aspects of the same phenomenon: electromagnetic interaction, which keeps electrons orbiting about the nucleus. Strong interactions are responsible for the cohesion of the constituents of an atomic nucleus; weak interactions, having a very low intensity and a very short range (less than one one-hundredth the size of an atomic nucleus), appear only when certain particles collide in certain nuclear reactions or disintegrations. We designate as "elementary particles" the smallest constituents of matter that can be isolated. These are categorized as a function of their interactions: the type of forces that they exert and to which they are sensitive. This model is remarkably successful when compared to the experimental evidence. It allows us to observe and to manipulate these miniscule and elusive objects—elementary particles.

The *general* theory of relativity has provided the foundation for the standard model of cosmology, the Big Bang model discussed above. However, when we use this model in an attempt to approximate in thought the primordial explosion, we inevitably encounter the problem of quantification of the gravitational interaction that dominates the dynamic of the Universe. This is the problem of the marriage of general relativity and quantum theory, or stated in our terms, the problem of simultaneously taking into account G, h, and c. By adopting another approach, particle physics confronts the same problem: how can we continue to keep gravitational interaction separate from our attempts to unify fundamental interactions, while pretending that it is negligible on the elementary scale! It thus appears to be entirely possible that, on the ultramicroscopic scale, where quantum effects can no longer be neglected, the gravitational interaction has an intensity comparable to those of other interactions.

QUANTUM THEORY OF FIELDS h *AND* c *CONSIDERED SIMULTANEOUSLY*

New difficulties arise with the new domain that is opened by considering the two constants h and c which, in this case, are not a "naturally" linked pair. Such an association "produces" the "quantum field theory" which, roughly speaking, is similar to the theory of elementary particles. In describing the quantum of action, we referred to the famous Heisenberg inequalities which show that, unless we conduct an experiment lasting an infinite period of time, we cannot observe a particle without even slightly modifying its energy. However, these Heisenberg inequalities can be interpreted in another way: when we wish to explore a structure with a great deal of temporal or spatial precision, it is necessary to impart to it a very high energy. But the special theory of relativity tells us that energy of motion can be transformed into energy of mass through the production of new particles. The energy that must be added so a structure with a high degree of spatio-temporal resolution can be explored will thus not only cause particles to move, but also create new ones! Elementary particle physics, which attempts to determine the ultimate constituents of matter, thus appears to be a physics where the number of particles is not conserved. In this sense, it is more like chemistry than like billiard ball mechanics; it involves "open" physics where all the reactions between elementary particles are connected to each other, so we must consider that it deals with quantum systems with infinite degrees of freedom. Such systems are known as *quantum fields*.

To describe the essential concept of the field, first consider the way it is used in *classical* electromagnetism. Coulomb's Law, named for the French physicist Charles Coulomb (1736–1806), defines the force between two electric charges. Let us assume that a group of point charges is spread throughout space. We can describe the forces exerted by this group of charges on a test charge located at any point in space. The electric field will then be defined as the vector sum at each point in space of all the forces exerted by point charges of the group. The value resultant of the electric field, that is, the resultant at each point in space of the electric force exerted on a test charge, is completely equivalent to the resultant of the charges of the group. When we wanted to consider relativistic effects in electromagnetism, it was necessary to define, in the same way, an *electromagnetic* field (with its electric and its magnetic components) as an infinite structure spread out over all *space-time*. If we consider not only c but also h, that is, if we wish to acknowledge quantum effects, we arrive at the extraordinarily rich concept of the "quantum field."

THE FEYNMAN PATH INTEGRAL

The theory of elementary particles and the interactions in which they participate consists essentially in modeling the probability amplitudes of all the possible reactions among elementary particles. This involves a program that may appear to be excessively ambitious; nevertheless, its execution has, for the most part, been undertaken and completed with remarkable success in certain cases. The source of all progress made in this area is the method of quantification proposed by the Nobel Prize winner Richard Feynman and called the path integral method. Unfortunately, to be able to fully appreciate the value of this method, it is necessary to have some familiarity with the technical tools of quantum theory. It is beyond the scope of this short book to go into these details. However, I would like to discuss the significance of this method and, as I did in the previous chapter, use the paradigm of the horizon and the real-potential-actual articulation.

The Feynman path integral corresponds to what I called above “the thinking of the world as the place of all possible horizon lines.” Let us consider, Feynman suggests, a certain transition of a system from an initial state to a final state. To find the probability amplitude for this transition, we first attempt to determine all the indiscernible paths that this transition can take, then write its probability amplitude as the complex sum of all the probability amplitudes associated with each of these indiscernible paths. Such a step has much in common with the one described by François Jacob in *Le Jeu des possibles*: “The scientific approach relentlessly confronts what could be and what is. This is how we can construct a representation of the world that is always closer and closer to what we call reality."* The philosophical merit of the Feynman method should be fully appreciated for its true worth which, in my opinion, is immense.

The path integral method avoids the impasse of the philosophical confrontation between subject and

* François Jacob, *Le Jeu des possibles* [The interplay of the possible], Paris, Fayard, 1983, p. 30.

object. Here the microphysical real is considered to be like the pole of a triad whose two other poles are the potential and the actual. The real is undeterminable *a priori*; it is not the subject of any criterion nor of any hypothesis *a priori*, with the obvious exception of the hypothesis of its existence independent of thought. In an attempt to determine, *a posteriori*, this real, we begin by envisioning all its possible forms, which make up what we call the *potential* pole of the triad. This potentiality space, which encompasses all the imaginable forms of existence of the real, or all the indiscernible transition paths, is determined by the properties of symmetry that are equivalent to the *laws of conservation*. These laws define the dynamic variables (energy, momentum, quantum numbers for particles identifiable in detectors), that is, the attributes of the *actual* pole of the triad.

THE PARADIGM OF QUANTUM ELECTRODYNAMICS

The first concrete result of the Feynman path integral method is the construction of a quantum theory for the electromagnetic interaction. By using the classical approximation, the Maxwell equations that describe this interaction can be deduced from a least-action principle formulated on the basis of certain properties of symmetry. In representing the electromagnetic interaction, rather than considering the electromagnetic field, which in itself would be sufficient, it is convenient to consider the "electromagnetic potential." This potential is a group of four functions defined at each point of space-time, from which the field is derived (the components of the electric and the magnetic fields are obtained from the derivatives of the components of the potential). The advantage of this presentation is not merely academic; in my opinion it allows a certain latitude in the way the electromagnetic interaction is represented. Even in classical terms, this latitude proves to be very useful for quantification. For a given electromagnetic field, there is, in fact, an infinite class of electromagnetic potentials from which this field is derived. Moreover, I find the term "potential" to be particularly well-suited, because if we consider the electromagnetic field, which is experimentally measurable as belonging to the category of actuality, the indetermination of the "potential" suggests, in the true meaning of the word, that it belongs to the category of *potentiality*. In any case, this indetermination of the potential is what is known as a property of symmetry (also known as *invariance*) of the electromagnetic interaction: the interaction is invariant by transformation from one potential to another giving rise to the same electromagnetic field. Such a transformation is known as a *gauge* transformation, and the invariance to which it corresponds is known as *gauge invariance*.

When an electromagnetic interaction is to be quantified, it is appropriate first to quantically treat the interacting matter. Consider the most simple example where the field of matter is a quantum electron field. In the wave representation, this field is a probability amplitude, dependent on space and on time, for which the modulus squared is the probability of finding an electron at each point of space-time. Let us remember that only the modulus of a probability amplitude can be measured, whereas its phase is indeterminate. Quantum electrodynamics thus acquires a new property of symmetry or of invariance, specifically quantic: the *invariance under change, at a constant modulus, of the phase change of the electron field*. This phase invariance is equivalent to the law of conservation of electric charge for electrons. This causes a "miracle" to occur: this phase invariance under change of the electron field can be ascribed to a *local* invariance, that is, as a phase-change invariance, *dependent on the point in space-time where it occurs*, if and only if the quantum field of electrons is coupled to a quantum field of interaction obeying the same gauge invariance as the classical electromagnetic potential. Thus it appears that the "horizontal" invariances, which are, on the one hand, the phase invariance of the quantum field of matter and, on the other, the gauge invariance of the quantum field of interaction, are both perfectly adequate to one another. This results in the elaboration of the theory of *quantum electrodynamics*, which is said to be a "gauge invariance theory."

Moreover, the miracle mentioned above appears to have an implication that goes beyond the scope of the electromagnetic interaction and concerns the three other fundamental interactions—the gravitational interaction and the two interactions with microscopic ranges, namely, the strong nuclear interaction and the weak nuclear interaction. The gravitational interaction, described on the classical level by the theory of general relativity, is, in effect, invariant with regard to local change of the reference frame, and this *locality* of the invariance, which we emphasized in Chapter I, has some similarities to the locality of the invariance of the electromagnetic interaction gauge. In fact, in the 1920s, Hermann Weyl (1885–1955) proposed an alternative to the geometric theory of quantum electrodynamics, similar to general relativity; it was on the occasion of this attempt that he introduced the concept of gauge invariance to which his name was later given. However, this attempt was abandoned, because quantum theory was not yet sufficiently developed. Moreover, the quantification of general relativity raised other difficult questions to which we will return later. Nevertheless, quantum electrodynamics played the role of an authentic paradigm in the elaboration of theories for nuclear interactions. For the two nuclear interactions, there is actually a local horizontal symmetry for the fields of matter which articulates a symmetry for the field interaction gauge. It is therefore reasonable to test theories of gauge invariance for these two interactions.

The construction of a coherent quantum theory for fields, and its application to the description of electromagnetic interactions, weak nuclear interactions, and strong nuclear interactions, was in the works for some fifty years. The first accomplishment was the elaboration of quantum electrodynamics (abbreviated as QED), which is the quantum and relativistic theory of electromagnetic interactions.

This theory has served as the testing ground for the entire conceptual construct of the quantum theory of fields, which allowed for concrete quantitative progress in the path integral method (Feynman diagrams, perturbation development, radiative correction, renormalization procedure, etc.).

We have thus succeeded in finding that the effects of quantification are capable both of being calculated theoretically and measured experimentally. The comparison of theory to experiment has proved extremely satisfactory. The "magnetic moment of the electron," which, in the classical approximation, is equal to 2, was measured to be 2.00231930482 ± (40) (the error relates to the last two digits); the predicted value is 2.00231930476 ± (52) (again, the error concerns the last two decimal places). It has been said that this agreement of theory and experiment is the best ever obtained in the history of science!

FROM QUARKS TO THE STANDARD MODEL

Quantum electrodynamics thus became the reference theory for the model on which the theories of nuclear interactions are based. It was first necessary to collect experimental data on these nuclear interactions and to construct a phenomenology. During this data collection phase, which took place for the most part during the 1950s and 1960s, a family of particles known as *hadrons* were discovered. Like protons, they can participate in fundamental interactions, and thus in the strong nuclear interaction in particular. During the 1960s, this family of hadrons proliferated to the point where the elementary nature of all its members began to appear doubtful: how could we imagine that the number of fundamental building blocks of matter would approach three hundred?!

The decisive turning point in elementary particle physics and fundamental interactions was the demonstration of a *new level of elementariness*, below the hadron level, that can be called *sub-hadronic*. The elementary particles of this sub-hadronic world are named *quarks* (after the characters in

James Joyce's *Finnegans Wake* who travel in groups of three; the proton is composed of three "quarks").

This word is well chosen to captivate the mind and maintain an air of mystery and curiosity about the reality of its subject. The quark scale represents what is today considered the ultimate level in the chain of the infinitely small. Protons, neutrons, and, more generally, all hadrons are composed of quarks. When the subject was first discussed during the 1960s, as a result of the work of Murray Gell-Mann in 1964, quarks were presented as a simple, purely "abstract" mathematical model. However, little by little, these particles acquired a more substantial and precise status. Nevertheless, they still raise a serious question: they exhibit the bizarre property of being free only within hadrons. This means that they cannot be isolated outside hadrons. The 1990 Nobel Prize was given for work—known as the Stanford experiment—which demonstrated the granular structure of protons and neutrons, thus confirming the validity of the quark model. However, quarks have never been found outside hadrons!

The theory of strong nuclear interactions at the quark level constitutes what is known as *quantum chromodynamics* (abbreviated as QCD). This theory has existed for some fifteen years, and is for the most part well understood by physicists. It is so named because it deals with "color" ("chromos" in

Greek) of quarks and, moreover, closely resembles quantum electrodynamics. Just as electrons interact by exchanging photons, quarks interact by exchanging bosons—known as "gluons"—which affect the color of quarks.

I believe that the terminology of color expresses the significance of the horizon paradigm. Indeed, we can say that a horizon line separates the hadronic world from the subhadronic. If the hadronic world is on our side of that line, because we can experimentally detect hadrons, we can then imagine that quarks and gluons, which have never been detected directly, are located beyond the horizon. How can a terminology be found more appropriate than that of color, which would be reserved for the subhadronic world, while the hadronic world would only be black and white?

The construction of the theory of weak nuclear interactions has some similarities to the theory we just described. In addition to quarks, which participate in all interactions, there is a family of particles of matter known as *leptons* (which includes the electron) that participates in the weak interactions but not in the strong interactions. Quarks participate in weak interactions by means of another one of their characteristics, which we call *flavor* (oddly enough, the word "quark" means "cottage cheese" in German!). Continuing with the analogy between electrodynamics and chromodynamics interactions, quarks interact in weak interactions by exchanging *intermediate bosons* which act on their flavors. For leptons, the equivalent of the flavor of quarks is known as "weak isospin" (or isotopic spin).

For more than twenty years, the elaboration of a theory for the weak interactions was focused on the question of intermediate bosons, first with the hypothesis of their existence, then with their experimental discovery in 1983 at CERN, the European Nuclear Research Center near Geneva, as the result of an extraordinary scientific adventure that has been told many times, and now with their systematic study using the LEP Supercollider at CERN, the "intermediate boson factory." This machine, which began operation in 1989, has provided completely satisfactory results; all the predictions for the standard model (which combines quantum electrodynamics, quantum chromodynamics, and the weak interactions theory) were confirmed with a high degree of precision. In *Matter-Space-Time*, a book co-authored with Michel Spiro, I wrote about this model: "In the future, we may look back and say that this was a *true scientific revolution.*" Today, with the abundance of results from LEP, I believe we can say that the standard model does indeed constitute a scientific revolution, solidifying the marriage of special relativity and quantum theory.

COSMOLOGY AND THE UNIFICATION OF FUNDAMENTAL INTERACTIONS

In light of the success of this standard model, it is tempting to go beyond the analogy and attempt to *unify* fundamental interactions, that is, to integrate them into a theoretical synthesis of the whole, just as the Maxwell equations were used to unify electrical, magnetic, and optical phenomena. This search for unification, in fact, constantly guides the scientific approach, and the history of science attests to the fact that every scientific revolution has been accompanied by a certain degree of unification. The recent history of particle physics, which is at most twenty years old, demonstrates that the tendency to unification has, in this discipline, assumed a dominant aspect. Moreover, by proceeding in this way, particle physics has come spectacularly close to cosmology, another branch in the forefront of physics. The final portion of this book is devoted to describing this reconciliation.

To understand the way this unification approach works, it is necessary to use concepts that are unfortunately very difficult to explain: *renormalization* and *symmetry-breaking*. Without going into technical detail, I will attempt to describe the essential significance of these developments.

The renormalization procedure was developed in response to a technical problem that arose during the use of the path integral method. The development of a probability amplitude for a certain transition, as the sum of the amplitudes associated with all the indiscernible paths by which the transition under consideration may occur, involves the Feynman "diagrams" and "amplitudes." But it happens that certain amplitudes associated with certain paths can be calculated only from integrals that "diverge," that is, are equal to infinity. These divergences are related to the behavior of transition amplitudes at very high energy, that is, as a result of the Heisenberg inequalities, and to their behavior at extremely short distances. This is obviously a major problem, since an infinite probability is something that has never been observed! The renormalization procedure allows us to overcome this problem: the "infinite part" is removed from these integrals by introducing some *ad hoc* parameters to the theory. The value of these parameters can be determined experimentally. A theory is said to be *renormalizable* if the number of *ad hoc* parameters necessary for the elimination of all the possible divergences is not infinite. A renormalizable theory can thus be compared to experimental results; since it is refutable, it can then, according to Karl Popper, be considered to be a true scientific theory. It is for this reason that the criterion of renormalizability is considered to be a decisive one for theoretical particle physics. Now, a new miracle of gauge invariance is that the gauge invariance theories are *renormalizable*, whereas the previous attempts to interpret weak interactions as contact interactions were condemned to failure because they resulted in a theory that was not renormalizable. It was also found that the gauge invariance theory adapted to the weak interactions could very naturally be incorporated into a scheme capable of unifying this interaction with the electromagnetic interaction in the same "electroweak interaction."

The renormalizability of a theory results in an extremely interesting property, which, in my opinion, is not unrelated to the "horizontal" point of view that has been developed throughout this book. The elimination of divergences in the renormalization procedure consists in absorbing the infinities in a redefinition of the parameters on which the theory depends. For example, the constant that measures the intensity of the interaction, known as the coupling constant or the charge associated with the interaction, is a constant that varies depending on the precision with which the interaction is described. To use the horizon paradigm, we can say that the horizon was drawn with a precision beyond which the divergences are eliminated, and that the line of this horizon is modeled by using effective quantities, which depend on the observation conditions. Let us accept that, in a renormalizable theory, such as the gauge invariance theories used for electromagnetic and nuclear interactions, the coupling constants must be considered as effective constants, dependent on the precision, in this case the energy. The way in which these constants are dependent on energy is completely predicted by the theory. For electromagnetic interactions, the effect of renormalization is barely perceptible, since the charge on the electron is almost completely independent of the energy. This is not the case for strong interactions where the coupling constant, also known as the "color charge," decreases appreciably when the energy increases, so that at high energy the "strong" interaction is weak! According to the theory, it is even predicted that an enormous energy exists which is totally inaccessible to experiment, where the weak and the strong interactions have comparable intensities. Moreover, since, as we stated above, it seems possible to unify the electromagnetic and weak interactions, it is possible to envision a grandiose *electronuclear* unification.

The reconciliation with cosmology has resulted in a profound change in perspective. Astrophysics, as we have said, integrated the observational data for the expansion of the Universe into the cosmological model of the Big Bang. According to this model, the Universe has been expanding and cooling ever since the primordial explosion some 15 billion years ago. The temperature decrease of the Universe is inversely proportional to the square root of the time elapsed since the Big Bang. Since temperature is just the average kinetic energy of the particles constituting the primordial Universe, we can say that the standard cosmological model also has a time-energy relation, similar to the one deduced, in quantum theory, from the Heisenberg inequality. Henceforth, using a high-energy probe to analyze a structure with a high degree of spatiotemporal resolution is, in a certain sense, to *go back in time*—in an attempt to reproduce in the laboratory the conditions of the primordial Universe as it was when its temperature was the same as the energy of the probe. This is how particle physics acquired an amazing temporal dimension.

Since renormalization theory suggests that the unification of interactions occurs at very high energy, the image produced by the reconciliation of particle physics and cosmology is that of a Universe which is not only expanding and cooling, but also has been *becoming* and *evolving* to its current state by passing through a series of *phase transitions*, ever since a dense and hot phase when all the interacting particles were undifferentiated. It was during these phase transitions that the interactions became differentiated from one another and produced the successive integrations leading to the series of interlocking structures that we observe.

It is quite natural that in an attempt to understand the phenomena of the differentiation of interactions and the integration of complex structures, we turn to statistical physics, in which the phase transition theory is precisely one of the major objectives. Thus, the mechanism of *spontaneous symmetry breaking*, borrowed essentially from superconductivity physics, has become a constitutive element of the standard model for the electroweak interaction. This mechanism could account for the differentiation of the electromagnetic and weak interactions that occurred one billionth of a second after the Big Bang. This mechanism assumes the existence of a new quantum field known as the *Higgs field*. The minimal model of spontaneous breaking of the electroweak symmetry assumes the existence of at least one new particle, the *Higgs boson*. Giant machines scheduled to begin operation at the beginning of the 21st century—the European Supercollider (LHC) and the American Supercollider (SSC)—will be dedicated to the search for this new particle, or any other effect related to the breaking of the electroweak symmetry.

In our treatment of quantum chromodynamics and the quark model, we discussed a characteristic property of quarks and gluons: the impossibility of ever isolating them from the hadrons to which they belong. This property is known as *confinement*. The analogy to quantum electrodynamics strongly suggests that in quantum chromodynamics, there may exist a "deconfined" phase in which the quarks and the gluons are free, similar to the *plasma* phase of electrodynamics in which ions and electrons are free. The experimental study of this new state of matter, which is a *plasma of quarks and gluons*, also known as "quagma," straddles both particle physics and nuclear physics. It is hoped that by causing collisions between high-energy heavy nuclei, the conditions of density and temperature will allow a deconfinement transition to occur. The problem with this type of experiment, which results in spectacular events such as the production of thousands of particles in a single collision, resides in the extraction of significant information concerning a possible transition from extremely complex experimental data. The theoretical estimations concerning the conditions of density and temperature for this transition would, within the framework of primordial cosmology, assign the time of the formation of hadrons from previously free quarks and gluons as having occurred around one one-hundred thousandth of a second after the Big Bang. The theoretical framework for such estimates is formed by not only quantum and relativistic physics, but statistical physics as well, thus explicitly taking into account h, c, and k. This new framework, which has begun to be explored, is the quantum theory of fields at finite temperature or density.

THE HORIZON OF QUANTUM GRAVITATION

This brings us back to the essence of this work, the consideration of universal constants. It will not have gone unnoticed that one of the four fundamental interactions—gravitation—has so far been left aside. However, the study of gravitation, in the domain of particle physics, or that of primordial cosmology, assumes that the three universal constants h, c, and G—and perhaps even all four if k is included to account for the effects of finite temperature or density—are considered simultaneously. It is for this reason that quantum gravitation appears as a kind of ultimate challenge to the whole of physics.

Before going any further, it is appropriate to caution the reader to realize that the questions of quantum gravitation appear only on an extremely distant horizon. To gain some appreciation of the order of magnitude, it is always advisable to use units adapted for the purpose. In particle physics, the system of natural units is the one in which h* and c are exactly equal to 1. These units are natural in relativistic physics (the speeds of particles are very close to the speed of light) and quantum physics (the actions involved are always on the order of a quantum of action). In this system of units, everything

* This is actually h divided by 2π.

occurs as if there were only one fundamental physical quantity instead of three, because two universal constants involve fundamental quantities that are set equal to 1. We can select energy to be the fundamental quantity, and in that case length, time, and mass become derived quantities, expressed in terms of energy: mass has the dimensional content of energy; length and time have the dimensional content of the reciprocal of energy. The unit of energy used is the GeV, or gigaelectronvolt, which is the energy imparted to an electron by a potential difference of a billion volts. A mass equal to 1 GeV is approximately equal to the invariant mass of the proton (which exactly equals 0.938 GeV). A length equal to 1 GeV^{-1} equals a fraction of a *fermi* or 10^{-13} cm, a length comparable to a nuclear radius. A time equal to 1 GeV^{-1} equals about 10^{-25} seconds, a typical duration for a nuclear interaction, which is the time it takes light to travel one fermi.

The dimensional content of the gravitational constant is the product of action and speed divided by mass squared. In the system of natural units, G is approximately equal to 10^{-38} GeV^{-2}; gravitation can thus legitimately be neglected in particle physics. Nevertheless, the existence of the three universal constants implies the existence of a fundamental mass known as the Planck mass, which is equal to 10^{-19} GeV, a fundamental length known as the Planck length, equal to 10^{-19} GeV^{-1} or 10^{-33} cm, and a fundamental time called "the Planck time," equal to 10^{-19} GeV^{-1} or 10^{-44} seconds; these are scales of energy, length, and time, for which gravitation, necessarily quantic, can no longer be neglected.

Now the classical model of the Big Bang does not consider quantum effects in the equations for gravitation. The Big Bang is thus a singularity corresponding to an infinite value of temperature and density at time zero. But at the Planck time, or 10^{-44} seconds after the Big Bang, the classical model is certainly wrong. As we do not yet know how to quantically treat gravitation, we consider the Planck time to be a horizon of primordial cosmology. In short, we designate the state of the Universe at the Planck time as the "Big Bang," without necessarily hypothesizing an "original singularity."

The gravitational interaction is extraordinarily weak at the energies involved in particle physics. It is intense on the macroscopic scale only because its range is infinite and it is attractive for all the particles of which matter is composed (as opposed to the electromagnetic interaction, which is attractive or repulsive, depending on the sign of the electric charge). Although gravitation cannot be neglected in the macroscopic world and even seems to be dominant in the Universe on a very large scale, it appears to be completely negligible on the scale of elementary particles. Moreover, the hypothetical boson of quantum gravitation, the *graviton*, interacts so weakly that it has not yet been possible to demonstrate its existence experimentally. *Gravitational waves*, predicted by the general theory of relativity, have also not been demonstrated (their detection is one of the challenges to astrophysics in the 21st century).

But, you may ask, why should we be preoccupied with the problems of quantum gravitation? I could turn the question around: if you have already heard of the Big Bang, did you ever wonder about the significance of this mysterious "primordial explosion," or wonder what there was "before the Big Bang"? We have just said that what is known as the "Big Bang" is the state of the Universe at the Planck time, the time at which gravitation is quantic. We cannot concern ourselves with the enigmatic and fascinating dynamic of the Big Bang without being preoccupied with quantum gravitation. The new interpretation of the gravitational constant, when it is associated with h and c, opens up amazing prospects: thus the Planck time and length suggest a quantum structure of space-time itself. Imagine the fascinating implications of a limit to the divisibility of space and especially the *divisibility of time*!

I would like to include a few words here to suggest the intense intellectual turmoil that has been caused by the quantification of gravitation. If we wanted to generalize and apply the Feynman path integral method to gravitation, it would be necessary to sum over all the possible measures of space-time! The attempt has been made to formalize this question, but the obstacle encountered was the non-renormalizability of the theory. Nevertheless, progress has been made with what is called the "superstring" theory, which appears to be quite promising, regardless of the reservations that many physicists and astrophysicists may have with regard to the schemas which at present are subject to pure speculation. In this theory, fundamental objects are not punctual particles, but rather one-dimensional objects known as strings. Whereas punctual particles travel along the lines of the Universe (the "paths" of the Feynman integral), strings sweep a two-dimensional "surface of the Universe". To quantify string theory is equivalent to summing over all the possible surfaces of the universe. This is akin to a quantum gravitation problem, strictly speaking, a quantum gravitation problem but in two-dimensional space-time. It has been determined that in two-dimensional space-time, general relativity could be quantified in a *renormalizable* theory!

String theory was first formulated at the beginning of the 1970s in response to the needs of the phenomenology of strong interactions. This model is designed to explain the confinement of quarks, which appear as the ends of strings or magnetic poles. We have seen that they cannot be isolated, no more than the end of a string can be isolated when it is cut. In this case we really have two pieces of string!

However, the status of string theory changed once it became known that it could play another role: that of an approximation to quantum gravitation involving the Planck length as a fundamental length. This revolution, which was achieved by our late colleague and friend Joël Scherk, led to an extraordinary result with the explosion of "superstrings" in 1984, when it was believed that a criterion allowing a completely constrained superstring theory to be selected had been discovered. This theory would be capable of unifying all the fundamental interactions, including gravitation, in a 10-dimensional space-time in which six dimensions are folded back on themselves.

Since then, we have had to become less ambitious, because it has been found that the criterion mentioned above is not as constraining as previously believed. Investigations are nevertheless continuing unabated: invariance *conforms*, which is the property of "horizontal symmetry" of superstring theories. It has an exceptionally rich conceptual content, encompassing and generalizing gauge invariance. Thanks to the *supersymmetry* that connects bosons and fermions, it is becoming possible to unify matter and interactions more profoundly than ever before. The fact that the string theories emerged from the physics of strong interactions makes me think that the "electroweak paradigm," which is starting to outlive its usefulness, can be extended by a "strong-gravity paradigm," thereby opening a new road toward the unification of fundamental interactions.

TOWARD A SCIENTIFIC COSMOGONY?

Particle theoreticians thus became convinced that they had to enlist the quantification of gravitation to go beyond the standard model of particle physics. It appears that this is also the key to extending the standard model of cosmology. The classic Big Bang model actually has some problems that some have suggested could be solved by imagining an exponential expansion phase known as *inflation* in the primordial Universe, or theory of *inflationary universe cosmology.* According to cosmologists such as Andrei Linde, inflation related to quantum gravitation and occurring at Planck time could lead to resolving the difficulties of the standard Big Bang model while retaining its wealth of acquired knowledge. Another school of thought, led by Ilya Prigogine in Brussels, interprets the Big Bang not as a singularity, but rather as a *quantum instability of space-time*. At the Planck time, flat space-time, devoid of all matter, is believed to have been quantically unstable and to have collapsed by becoming curved and producing all the matter contained in primordial entropy in the form of three quantum black holes.

Such scenarios are fascinating, but they cause serious reservations on the part of many scientists who would prefer to exclude cosmogony from the field of science. These reservations are also fueled by what is called the *strong anthropic principle:* "The Universe must be such that it allows the creation of observers in its midst at any stage in its evolution" (B. Carter), a statement which actually has a "strong" teleological content and provokes religious speculation!

However, in conclusion, I would like to give my personal position on this debate. I believe that one of the roles of science is to shed new and relevant light on the cosmogonic debate. Science is a part of culture, and basic research attempts to bring together the elements of the answers to the philosophical questions that have preoccupied humanity since the dawn of time. But it is precisely with regard to such matters that the attitude of humility I referred to at the beginning of the book becomes meaningful: it would be tragic if science, under the pretext of being interested in questions relating the origin of the Universe, were to serve as justification for normative, obscurantist, or scientistic pseudophilosophies.

After having been formulated as strictly utilitarian, the universal constants that we know could be accepted as founded in nature—we could speak of them as "constants of nature." Today, we are compelled to return to them, but in a new sense, with a utilitarian interpretation: they are not physical constants of the Universe, but rather universal constants of physics; they express a "self-discipline" necessary to how we think of their relationship to nature. We can consider them to be "safeguards" that physics imposes on itself. In the first place, this is the case for c, the upper limit for all velocities, which expresses the impossibility of immediate action at a distance: it is, after all, completely reasonable to acknowledge this impossibility. We saw how Einstein was able to proceed with extraordinarily bold challenges because he forbade himself to even consider that any speed could exceed that of light. Boltzmann's constant and Planck's constant also express impossibilities, limits, and constraints which would be unreasonable to deny: all knowledge comes at a cost; no perpetual motion machine can exist; it is impossible to influence the past. It is by fully accepting these constraints that quantum theory has allowed so many scientific and technological advances to have been made in the 20th century.

At first sight, Newton's constant G does not appear capable of being interpreted as expressing a certain impossibility. On the contrary, the theory of gravitation developed by Newton appears to mark the release and liberation from anthropocentric illusions, a window on the global thinking of the Universe. Nevertheless, we have seen that, when it is associated with h and c, Newton's constant can be interpreted as a limit to the divisibility of space and time. This limit is so distant, however, that it appears to be only slightly constraining. But what a stimulus to the imagination of physicists, providing they do not abandon reason!

This is the lesson of this book: that the universal constants, our constants, not only keep us from wandering, but also open horizons to us. If other constants should appear, we already know that they will destroy physics as we know it. They will take their place alongside our four very solid guideposts to shape our approach to the exploration of nature. Other "safeguards" may appear, but they will also open new horizons that are continually retreating.

BIBLIOGRAPHY

BATON, J. P., and COHEN-TANNOUDJI, G., *L'Horizon des particles. Complexité et élémentarité dans l'Univers quantique* [The horizon of particles. Complexity and elementariness in the quantum Universe], Paris, Gallimard, 1989.

BOUDENOT, J.-C., *Électromagnétique et gravitation relativistes* [Electromagnetism and relativistic gravitation], Paris, Ellipses, 1989.

BRILLOUIN, L., *Science and Information Theory*, New York, Academic Press, 1962.

COHEN-TANNOUDJI, G., and SPIRO, M., *La Matière-Espace-Temps. La logique des particles élémentaires* [Matter-Space-Time. The logic of elementary particles], Paris, Fayard, 1986.

FEYNMAN, R., *Lumière et Matière: une étrange histoire* [Light and matter: A strange story], Paris, Interéditions, 1987.

HAWKING, S., *A Brief History of Time. From the Big Bang to Black Holes,* New York, Bantam Books, 1988.

KASTLER, A., *Cette étrange matière* [This strange matter], Paris, Stock, 1976.

NEWTON, I., *Principia Mathematica*, trans. F. Cajori, Berkeley, University of California Press, 1962.

PRIGOGINE, I., and STENGERS, I., *Entre le Temps et l'Éternité* [Between time and eternity], Paris, Fayard, 1988.

REEVES, H., *L'Heure de s'enivrer. L'Univers a-t-il un sens?* [Time to rejoice. Does the Universe have meaning?], Paris, Le Seuil, 1986.

X'THUAN, Trinh, *La Mélodie secrète: Et l'homme créa l'Univers* [The secret melody: and man created the Universe], Paris, Fayard, 1988.

WEISSKOPF, V., *La Révolution des quanta* [The quantum revolution], Paris, Hachette, (*Questions de science* Series), 1989.